The Electrical Safety Program Book

The Electrical Safety Program Book

Kenneth G. Mastrullo

Ray A. Jones

Jane G. Jones

National Fire Protection Association
Quincy, Massachusetts

Product Manager: Brad Gray
Editorial-Production Services: Omegatype Typography, Inc.
Composition: Omegatype Typography, Inc.
Cover Design: Cameron, Inc.
Manufacturing Manager: Ellen Glisker
Printer: R. R. Donnelley/Willard

Copyright 2003
National Fire Protection Association, Inc.
One Batterymarch Park
Quincy, Massachusetts 02169

Notice Concerning Liability: Publication of this work is for the purpose of circulating information and opinion among those concerned for fire and electrical safety and related subjects. While every effort has been made to achieve a work of high quality, neither the NFPA nor the authors and contributors to this work guarantee the accuracy or complete-ness of or assume any liability in connection with the information and opinions contained in this work. The NFPA and the authors and contributors shall in no event be liable for any personal injury, property, or other damages of any nature whatsoever, whether spe-cial, indirect, consequential, or compensatory, directly or indirectly resulting from the publication, use of, or reliance upon this work.

This work is published with the understanding that the NFPA and the authors and contributors to this work are supplying information and opinion but are not attempting to render engineering or other professional services. If such services are required, the assis-tance of an appropriate professional should be sought.

NFPA No.: ESM03
ISBN: 0-87765-580-4
Library of Congress Control No.: 2003102869

Printed in the United States of America
05   06   07   5   4   3

Contents

CHAPTER 3

SETTING UP THE ELECTRICAL SAFETY PROGRAM   15

CHAPTER 4

PROCEDURES AND PLANS   25

CHAPTER 5

SITE ASSESSMENT   37

■ CHAPTER 6

HAZARD BOUNDARIES   45

■ CHAPTER 7

HAZARD/RISK ANALYSIS   57

■ CHAPTER 8

PERSONAL PROTECTIVE EQUIPMENT   69

■ CHAPTER 9

LOCKOUT/TAGOUT   79

■ CHAPTER 10

TRAINING   91

■ APPENDIX K
BUDGET SAMPLES AND CHECKLISTS   253

Preface

The goal of *The Electrical Safety Program Book* is to provide a framework for converting OSHA regulations and electrical safety requirements into a practical and workable plan. Electrical safety is everyone's business because it affects owners, supervisory personnel, workers, and families. By developing an electrical safety program, employers become the catalysts for encouraging and promoting an atmosphere of safety. A properly executed electrical safety program is an investment that pays great dividends financially, politically, and socially.

As the publisher of NFPA 70E, *Standard for Electrical Safety Requirements for Employee Workplaces*, NFPA has over a quarter century of commitment to safeguarding employees in the workplace. This book references the electrical requirements in NFPA 70E, which serve as the foundation of an electrical safety program in the workplace. The person responsible for the electrical safety program, regardless of his or her position, will find the logical presentation of information in this book to be a valuable resource.

The Electrical Safety Program Book has a chronological progression beginning with the history and development of electrical codes, standards, and regulations. It continues on the path of development of an electrical safety program from conception through finalization of a comprehensive plan that is tailored to a company's unique environment. The subject matter covered includes site assessment, procedures and plans, hazard risk analysis, personal protective equipment, training, budgeting, auditing and record keeping, and program administration. Many forms and checklists were created especially for this book that could be invaluable to the user. They include a site assessment form, energized electrical work permit, job briefing form, and task assessments in compliance with NFPA 70E.

Until now, there has been no comprehensive source of information that incorporates the OSHA regulations, electrical safety requirements, and the framework for an electrical safety program. It is the authors' hope that this manual will fill that void.

I would like to thank all the members of the electrical industry whom I have worked with and learned from. They have influenced many of the concepts in this book. In addition, I am grateful to my wife Dottie for her years of support and understanding and for being my guiding light. I also want to thank my son Brian and daughter Jeanine, who are the pride of my life.

Kenneth G. Mastrullo

How to Use This Book

This book has three distinct parts. Thirteen chapters provide discussion about various aspects of an electrical safety program. Eleven appendices contain illustrations of various components of an effective electrical safety program. A compact disc provides forms and templates from the appendices and a training program, all in electronic form.

Users should first read and study the discussion chapters to develop an understanding of some of the basic concepts that an effective electrical safety program contains. Next, the user should review the content of each appendix to understand what kind of assistance is available through the tools provided.

The information presented in the book is intended to be used as guidance only. The work environment is different for each corporation, employer, and location, and each has different needs. Each electrical safety program should be designed and implemented based on the working environment, as it exists. Language and phraseology vary substantially from one location to another or from one employer to another. All members of the work force must easily understand the content of the electrical safety program. Communication skills may be good at one location and not as good at another.

No electrical practice, procedure, or program should be implemented until after it has been reviewed in detail and its requirements determined to apply within the work environment. Although this book provides information that might be directly usable, change in the work environment could require significant changes. Before these or any other requirements are put into place, responsible authorities should determine that the requirements are adequate to address the concern.

The compact disc provides a draft component that can be used as the beginning point for a company to develop various aspects of its own electrical safety program. However, the authors urge caution when using program components from external resources, including this book. Each component of an electrical safety program should be tailored specifically to the needs of the company that will be using the program.

The Electrical Safety Program Book

Chapter 1

The Electrical Safety Journey

Thomas Edison and George Westinghouse disagreed about whether direct current or alternating current should become the "standard" distribution system for electrical energy in the United States. Each man saw advantages to one scheme that he considered to be superior. The alternating current distribution system envisioned by Westinghouse in the late 1880s would enable electrical energy to be transmitted long distances at higher voltages and then transformed to a lower voltage for use. Edison had no realistic vision for distributing electrical energy.

Both Edison and Westinghouse knew that electricity was dangerous. In the late 1800s, officials of the state of New York knew that electricity was so dangerous that they decided to design and install an electric chair for use in their prison system. Edison was approached and asked to design the chair. As Edison was not a proponent of capital punishment, he initially refused the contract. However, he recognized that the state probably would approach Westinghouse for help with its objective. Edison expected that an electric chair designed by Westinghouse would use direct current as the source of energy. Edison did not want his favored direct current system to be used for that purpose, so he agreed to take the contract.

Taking the same approach that he expected Westinghouse to take, Edison designed and helped to construct the electric chair that was used by the state of New York for many years. Edison's design used alternating current as the source of energy to prove that alternating current was more deadly than his favored direct current.

The distribution system that Westinghouse favored survived the disagreement. The ease of changing voltage served as the winning factor. Electrical energy could be transmitted over long distances at high voltage, which minimized line losses, and then it could be transformed to lower voltage for use.

Although it was known that electrical current flowing through a person's body was very unpleasant, no one realized the potential impact of the unpleasant feeling. Over the course of many years, people began to understand that the expanding use of electrical energy presented hazards that were very serious.

■ IN THE BEGINNING

Fire was the first hazard that was generally recognized. Electrical energy was initially used to supply lighting, and rudimentary electrically powered equipment was installed in major cities across the northeastern United States. The number of fires increased dramatically. Few people had the experience and knowledge of how to install electrical circuits and equipment to mitigate the fire hazard. Equipment and installations varied from one installation to another, and no codes or standards existed. Insurance companies bore the brunt of the economic losses that were attributed to the fires.

Representatives of several insurance, electrical, architectural, and other allied interests came together in 1897 and determined that an installation code was one key to minimizing exposure to the increasing number of fires. The outcome of that meeting was that an installation code would be drafted. The new code, known as the National Electrical Code, would establish requirements that would apply broadly in the electrical and architectural industry.

The National Fire Protection Association assumed responsibility for the National Electrical Code in 1911. Since that time, the NFPA has fulfilled the responsibility by generating, reviewing, and publishing the *National Electrical Code*® (*NEC*®). The initial purpose of the *NEC* was to minimize the use of electricity as the cause of fires in various facilities.

Following the publication of the *National Electrical Code* in 1913, the Institute of Electrical and Electronics Engineers published another installation code, called the *National Electrical Safety Code* (*NESC*). Although the *NESC* contained information related to facilities and architectural installations, its essential focus was related to transmission and distribution of electrical energy.

As the use of electrical energy expanded, little additional knowledge about electrical safety was available for many years. When the Wiggington solenoid voltage tester was developed and made generally available, workers had a device that could detect whether or not voltage was present before contacting a circuit component. The solenoid-type voltage tester was the next important electrical safety development. However, well into the 1960s, electricians were taught to determine the absence of voltage by keeping one hand in a pocket and touching a conductor with the thumb and forefinger of the other hand.

Eliminating electrocutions became an objective with the advent of the OSH Act of 1970. Electrical safety issues became a real concern among affected employers when the OSH Act was signed into law. With that act, associated economic penalties became a motivating factor for employers. Employers and corporations began to pay attention to electrical safety issues because of the authorized enforcement mechanics contained within the OSH Act.

Formalizing personal safety requirements served to raise the electrical safety bar. Within a few years after the OSH Act was signed into law, some employers and companies began to learn that electrical safety was good business. Insurance rates were reduced as the number of injuries was reduced, and legal costs were commensurately reduced. Production losses also were reduced. The return on investment in a safety program yielded a very substantial amount of money. Most managers could understand that result.

The rate of understanding accelerated with the knowledge that an investment in a safety program is good business. The milestones shown by the time line in Figure 1-1 illustrate the increased rate in developing knowledge and understanding associated with electrical safety.

■ CURRENT STATE

Three very important milestone events occurred very closely to one another around 1980. First, in 1979 (after the Occupational Safety and Health Administration requested it), NFPA published a standard that was intended to identify safe work practices. This was the first consensus standard that defined electrical safety-related work practices. The standard was NFPA 70E, *Electrical Safety Requirements for Employee Workplaces*. The second event occurred when Dr. Raphael Lee opened the first burn center related to injuries from electrical incidents at the University of Chicago. The University of Chicago Electrical Trauma Center leads the medical community in understanding how to treat victims of electrical burns. The third event occurred when R. H. Lee (Ralph Lee—no relation to Raphael Lee) authored a technical paper entitled "The Other Electrical Hazard,"

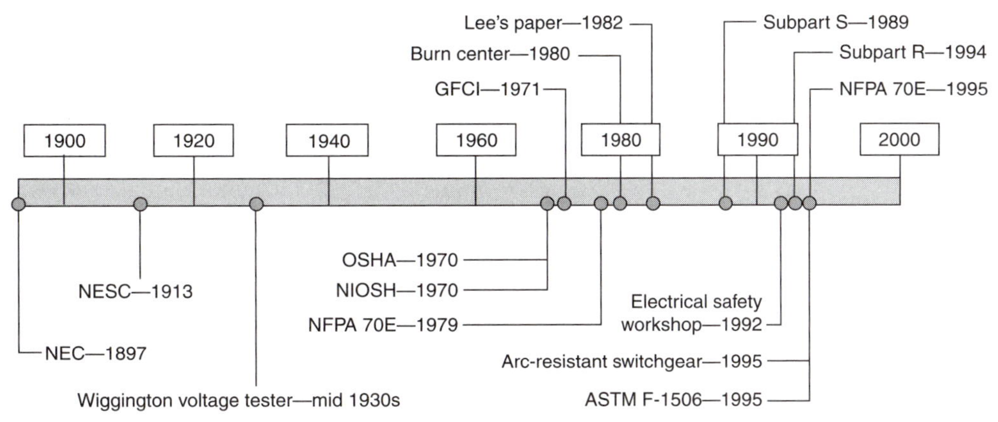

FIGURE 1-1 *Electrical Safety Journey over the Past 100 Years*

which was presented at an IEEE Petroleum and Chemical Industry Conference in 1982. These three events served to initiate an escalation in developing understanding of electrical hazards. As illustrated in Figure 1-1, the rate of knowledge generation has increased dramatically since those events.

With the publication of the 1995 edition of NFPA 70E, arc flash was recognized as an electrical hazard for the first time in a national consensus document. Since that time, several ASTM standards have been published that define testing requirements to determine the integrity of protective clothing when exposed to an electrical arc flash. With the publication of the 2000 edition of NFPA 70E, a broad consensus was established that determining incident energy is an acceptable way to categorize the degree of hazard associated with an arcing fault. Although NFPA 70E and several IEEE papers discuss methods to determine what protective equipment may be necessary to avoid injury from an arcing fault, only the thermal aspects of the electrical arc have been addressed. Other hazards, such as arc blast and flying parts and pieces, have not yet been investigated.

■ BUILDING AN EFFECTIVE SAFETY CULTURE

Organizational experts advise that a group of people functioning as a team is more effective than the same people functioning individually: the whole is greater than the sum of its parts. Experience supports that premise. A team of people working together can be a powerful force. The focused efforts of a team can influence how workers feel about the safety program in general and the electrical safety program specifically.

An employer should charter an electrical safety team (EST). The EST should report to senior management either directly or through the safety manager. Although members of the EST might be selected from the line organization, each member should be encouraged to express views and ideas freely. Managers and supervisors should express strong support for the EST and for its decisions. The EST should be encouraged to make technically sound choices. In addition, the EST should express strong support and encouragement for the safety efforts of colleagues.

Establishing a Goal

As the leader of a sports team, the coach always ensures that his or her team has both short-term and long-term goals. Normally, a short-term goal is to win the next game or to demonstrate improvement. The same team might have a long-term goal to win the division or world championship. Some sports teams seem to have developed a "winning attitude" where the team expects to win the next game and even to be conference champs. Many other attributes are necessary for a sports team to be winners. A long-term goal can be achieved by accomplishing all of the short-term goals.

The EST should establish a long-term goal. The authors suggest that the EST should consider setting the long-term goal to be zero injuries. Before selecting a goal, the EST should discuss the implications of the goal and consider how the workers would embrace it. Once a goal is identified, the EST should determine the mechanics of how to achieve it. All members of the organization should be convinced that the goal is achievable and that a realistic plan exists to enable reaching the goal. Members of the organization must be convinced that they have a critical and integral role in reaching the goal. The EST can identify milestones—points along the journey toward the long-term goal where celebrations can be held. Each celebration will encourage workers that progress is being achieved.

Establishing a Bias

Most electricians and electrical technicians have been trained throughout their careers to believe that keeping the equipment energized is a good thing. Workers are sometimes rewarded for performing an electrical work task with the circuit still energized. A worker might proudly state that he or she has had a little shock, but the equipment did not have to be deenergized. The culture in that organization has a bias toward performing work on or near electrical circuits while they remain energized. In some instances, supervisors encourage workers to accept the risk of injury.

One of the first objectives for an EST should be to shift the bias from routinely accepting the risk of an electrical injury to viewing risk of an electrical injury as unacceptable. A worker might ask a production supervisor to deenergize selected equipment that has a bearing on the amount of product. Unless the production supervisor understands that a person will experience an elevated risk of injury, he or she has no realistic reason to accept or reject such a request. The EST, then, must establish a process to ensure that each supervisor understands the relationship between energized circuits and an elevated risk of injury. The EST might choose to initiate a process whereby a written permit indicates the elevated risk of injury and requires a supervisor's signature before the task is executed (see Appendix E, Energized Electrical Work Permit).

Benefits of a Positive Safety Culture

The electrical safety program should be one element of the overall safety program for a company or organization. Electrical hazards are somewhat different from other hazards associated with other forms of energy. The way that a person can be exposed to an electrical hazard is different from exposure to other hazards. Since exposure to electrical hazards is different from other energy sources, specific requirements must be different. An effective safety culture recognizes and accepts these differences.

Workers are eager to participate in the safety process when the culture welcomes and accepts their input. People like to feel as if they are part of an important process. If the safety culture defines unnecessary risks as unacceptable, worker participation in the safety program is more direct. The safety culture has a direct impact on the amount of pride a worker exhibits in his or her work process. A positive work environment tends to increase a worker's positive feelings about his or her job.

Who Are the Stakeholders?

Everyone has a stake in the safety culture. Every person performing a work activity on the site has an interest in the safety environment. Although participation might come at different points in the process, every member of the line organization is affected in the event of an injury. The lives of spouses and other family members are also affected if an injury occurs. Contractors and contract employees are exposed to the same hazards as anyone else performing work on the site.

Positive and Negative Factors That Influence the Culture

People tend to *listen* to what managers, supervisors, and safety professionals say. After hearing the rhetoric, most people then *watch* how those managers, supervisors, and safety professionals act. What a worker sees is what he or she accepts as real. The issue, then, is that members of the line management "walk the talk." Workers understand the possibility that safety directives are nothing more than rhetorical verbiage. However, if members of line management "walk the talk" and demonstrate that safety directives are much more than rhetoric, then workers know that a safety directive defines what is expected. Then the worker knows what to expect.

Enforcement mechanics and methods should be predetermined. Representatives of the entire company should be involved in the process to determine enforcement policies. Workers can then expect consistent application of published enforcement policies. Just as enforcement policies are important, so is a recognition and reward system. An effective safety program will reduce the number of injuries; therefore, the costs will be reduced. A recognition and reward program serves to create continued interest in the program. The monetary value of the reward does not seem to be a major factor. If a reward is selected and then highly emphasized within an organization or company, workers tend to covet the reward.

Facility Considerations

Workers should be able to expect that a facility meet the requirements of national consensus standards. In most instances, local building codes define those requirements. When the facility is first occupied, each building meets all requirements of the local building code. However, the initial integrity of installation must be maintained. If equipment, buildings, and services are permitted to deteriorate, workers will be exposed to hazards in a different way. The EST should take necessary steps to prevent any significant degradation of the facility or equipment.

Overcoming Communication Difficulties

One of the most difficult problems for an EST to overcome is the ability to communicate. Many injuries occur because communications among groups of people are incomplete or inaccurate. In some instances, the communication difficulty occurs when people involved in the overall job do not speak the same language. Sometimes the inability to communicate exists even when all concerned are speaking English. For instance, the term "hot work" might mean several things. Hot work might mean working directly on energized terminals, working in the vicinity of energized terminals, or working on thermally hot equipment. When workers are exposed, or potentially exposed, to an energized electrical conductor or circuit part, each person involved in the work task must understand both the requirements and the limits of his or her portion of the overall job.

■ WHO IS AT RISK OF AN ELECTRICAL INJURY?

Every person who uses electrical energy for any purpose is exposed to risk of an electrical injury. The requirements of the installation code (the *NEC*) provide for adequate protection for people who use equipment or facilities. Workers in other interactions (such as construction or maintenance) with the equipment are not protected from injury by the Code. If electrical equipment is being maintained or if a worker is troubleshooting a problem, the operating conditions are *not normal*. If new equipment is being installed, conditions are *not normal*. During periods where these activities are performed, installation and building codes offer little protection from injury.

When a person is exposed to an electrical hazard, he or she is at risk of injury. A maintenance worker exposed to a shock hazard while troubleshooting with a voltmeter has the same exposure as a construction worker who is measuring voltage to determine if an energized condition exists. The risk and exposure are identical. Similarly, a utility worker performing a comparable work task also has exactly the same risk of injury. The risk of injury is, therefore, related to the hazard and how a worker is exposed to the hazard instead of to the industrial segment that employs the worker. Equipment operators who open a door to reset an overload relay also have similar risk of injury. Risk of injury is not related to work assignment.

How Many People Are Injured?

The affects of electrical incidents and their results are devastating. In the United States, 4,000 non-disabling and 3,600 disabling electrical contact injuries occur in the workplace annually. Several negative results occur when a worker is injured. A non-disabling injury is defined as an injury requiring hospitalization and time lost from work. This type of injury equates to lost or reduced income and a traumatic effect on the victim's family. The worker could be out of work for a couple of days or months. The company loses a significant part of its team. The workers may lose confidence in the procedures and policies of the company. Typically, workers question how the worker was injured.

In addition to these statistics, over 2,000 workers are sent to burn centers with electrical burns every year. Some of these workers are able to return to work, while others become permanently disabled. Whether or not they return to work, they must contend with other issues. They might be permanently disfigured or require amputations to survive. Their recuperation process is riddled with excruciating pain and suffering. Their life is restricted, due to sensitivity to cold weather, mobility, or other physical barriers. Their recovery period is usually long, and several surgical procedures might be required in an attempt to repair their bodies.

In addition to all of these injuries, over 400 workers die from electrocution each year. This statistic does not seem dramatic until you equate that with the fact that at least one person dies from electrocution in the workplace every day, and these statistics do not even include electrical burn fatalities. Figure 1-2 exemplifies an even more startling statistic. Of all electrocutions, the 25- to 34-year-old group is the largest, at 35 percent. The three groups that encompass 75 percent of the fatalities are aged from 20 to 44. What happens to the families of these victims? Their lives are dramatically affected forever. A key source of financial and emotional stability for the family has been removed forever. The overwhelming majority of these workers are in the prime of their lives. The elements of the graph can be used to address root causes of some incidents, such as job task assignments, knowledge, working procedures, training, and job culture.

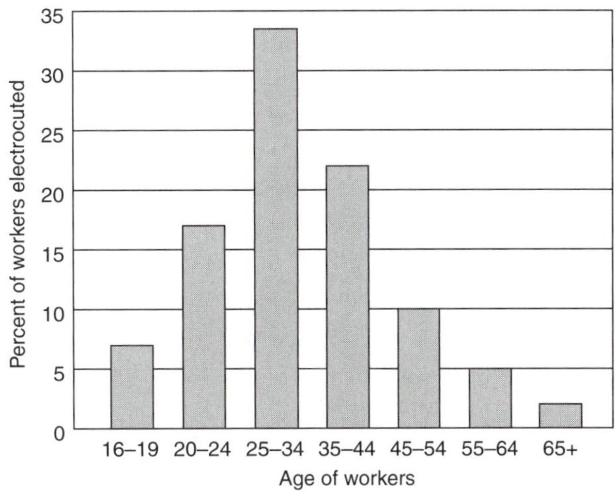

FIGURE 1-2 *Worker Deaths by Electrocution, 1982–1994*
Source: Worker Deaths by Electrocution: A Summary of Surveillance and Investigative Case Reports. NIOSH, 1998, Figure 5, p. 14.

■ CAUSES OF INCIDENTS AND INJURIES

An incident might occur as the result of someone performing an unsafe act or using defective equipment. Analysis of injury and incident records suggests the following reasons for an incident:

- Some incidents result from unsafe equipment. Equipment was not installed correctly, improperly protected, or not maintained adequately, exposing people to unexpected hazardous conditions.
- Some incidents result from unsafe conditions. For example, an unsafe condition exists when an energized conductor that is normally covered is left uncovered, exposing an unsuspecting person to an electrical hazard.
- Workers perform an unsafe act. Most unsafe acts do not result in an incident or injury. However, unsafe acts are the most frequent cause of an incident and injury.

The following list identifies some of the reasons people perform an unsafe act:

- Failure to recognize the hazard
- Lack of necessary skill, knowledge, and ability
- Lack of necessary training and understanding
- Previous performance of the same task many times without incident
- Pressure to finish the task
- Performance of a familiar task without thinking
- Distraction due to a personal issue or another employee
- Lack of familiarity with some unique aspect of the equipment
- Incomplete or inaccurate reference drawings
- Failure to follow the company procedure

- Faulty test equipment (equipment not inspected)
- Lack of a job plan
- Lack of respect for task limits
- Selection of the wrong tool

Corporate culture has an impact on the attitude of supervisors, midlevel managers, and workers, although these same people establish the culture. In most instances, the actions and attitude of senior managers set the tone for the culture that is established within an organization. It is senior management that must allocate resources to develop and maintain a program. What is the company philosophy regarding a safe workplace as part of the company's mission? Does the company see the linkage between a safe workplace and profitability?

Employee safety should be an integral part of every organization's mission. Most companies describe themselves as a great place to work. Their self-description normally includes a working environment that is free of hazards. Because of the fact that electricity is virtually invisible, integrating appropriate respect for electrical hazards can be difficult. The use of electricity is viewed as a safe experience. In almost all instances, electricity provides energy to perform tasks and the user is not even aware of the energy source. However, hazards associated with electrical energy tend to cause more severe injuries than most other kinds of energy. For all of these reasons, electrical safety deserves special consideration.

REFERENCES

ASTM F1506, "Specification for Textile Materials for Wearing Apparel for Use by Electrical Workers Exposed to Momentary Electric Arc and Related Thermal Hazards." Philadelphia: American Society of Testing and Materials, 2002.

Lee, R. H. "The Other Electrical Hazard: Electric Arc Blast Burns." IEEE Transactions, Vol. IA-18, No. 3, May/June 1982.

National Electrical Code® (ANSI/NFPA 70). Quincy, MA: National Fire Protection Association, 2002.

National Electrical Safety Code (ANSI/IEE C2). New York: Institute of Electrical and Electronics Engineers, 2002.

NFPA 70E, *Standard for Electrical Safety Requirements for Employee Workplaces.* Quincy, MA: National Fire Protection Association, 2000.

OSH Act: Public Law 91-596. Ninety-First Congress of the United States of America, S2193, Dec. 29, 1970.

OSHA Regulations 29 CFR 1900 to 1910. Washington, DC: Occupational Safety and Health Administration, U.S. Department of Labor.

OSHA Regulations 29 CFR 1926. Washington, DC: Occupational Safety and Health Administration, U.S. Department of Labor.

Worker Deaths by Electrocution: A Summary of Surveillance and Investigative Case Reports. Cincinnati, OH: National Institute for Occupational Safety and Health, 1998.

Chapter 2

The Relationship of Regulations, Codes, and Standards

Processes used to develop codes and standards are viewed as somewhat mysterious. Reading the content of a code or standard sometimes lends credence to the mystery, because the text of a requirement might be confusing and require interpretation. Experts sometimes disagree about the intent of specific requirements. However, all codes and standards that are published by a standards-developing organization (SDO) within the framework of the American National Standards Institute have some common characteristics. When coupled with knowledge about the SDO that published the standard, the common characteristics can help a user to understand a code or a standard.

Regulations are a different type of document. Codes are generally intended for adoption by regulatory bodies, and then they become regulations. The *National Electrical Code®* (*NEC®*) is an example of a code that is frequently adopted by governmental bodies. Sometimes the *NEC* is adopted by a state legislature, sometimes it is adopted by county governments, and sometimes it is adopted by a city government. When a code is adopted by governmental action, the same action defines inspection and enforcement methods. Generally, the inspector serves to interpret requirements.

Some governmental bodies write their own standard by selecting a consensus standard, if one exists, and then selectively embracing some requirements from the standard and rejecting others. The OSHA rules that cover electrical hazards serve as an example. Parts I and II of NFPA 70E-2000 are the basis for the Code of Federal Regulations' 29 CFR 1910 Subpart S, while parts III and IV are not addressed within the OSHA rules.

The terms *guide, standard,* and *code* are commonly used to have a similar meaning for most standards-developing organizations. As the term suggests, a *guide* is not intended to define requirements. Instead, a guide is intended to provide suggestions and may contain more than one possible solution. A standard defines requirements. A standard might provide alternatives, but it intends to define all acceptable choices. A code is very similar to a standard except that the language contained in a code is selected based on a need for the code to be enforceable.

■ COMMON CHARACTERISTICS

The intent of any standard or code should be interpreted based on the mission or objective of the organization that produced the document. For instance, the Institute of Electrical and Electronics Engineers (IEEE) is a worldwide organization that focuses on the interests of electrical and electronic engineers. One of those interests is product interchangeability among manufacturers, industries, and other borders. Writing and publishing standards is a way that the IEEE achieves that objective. Because one mission or objective of the IEEE is product interchangeability, a standard is one major tool that helps to achieve the mission or objective.

All codes and standards have a scope or purpose statement. This statement defines the objective of the document. The scope or purpose statement defines the limit of the document's application. If the limits are readily understood, then where the standard applies is more easily understood. From the interpretation point of view, a scope or purpose statement defines the boundary of the document. Sometimes a scope includes an exclusionary statement. For example, Article 90.2 of the *NEC* includes an exclusionary statement covering installations that are under the exclusive control of an electric utility. To understand the requirements of a code or standard, an equivalent comprehensive understanding of the scope is imperative.

■ INTERACTION OF STANDARDS

SDOs typically produce three different types of documents: standards, guidelines, and recommended practices. Although each product document might be called by a different name, the products tend to fall into these general categories. A standard is intended always to be applied and uses mandatory language such as *shall* and *must* to define requirements. If words such as *should* or *may* are used in conjunction with an idea, the standard implies that the user has a choice in interpretation.

When a standard is embraced and implemented, the user body or unit also determines how the standard is to be interpreted and enforced. If the standard is implemented by regulation, the same regulation must also identify who is accountable for interpretation, inspection, and enforcement. The term *authority having jurisdiction* is frequently used in standards' language to mean "this accountable agency." It follows, then, that if a standard becomes legally binding, the regulation overrides any owner-enforcing authority. However, if an owner adopts a standard (where no regulation exists), then the owner must also identify the mechanism for inspection and enforcement.

One element of an electrical safety program should identify all standards that are embraced by the owner or employer. Standards that have been adopted by regulation also should be identified in the electrical safety program.

■ STANDARDS ACCEPTED BY THE AMERICAN NATIONAL STANDARDS INSTITUTE

The American National Standards Institute (ANSI) does not develop standards. Instead, ANSI monitors the development of standards by other organizations. ANSI focuses on *how* the standard was developed. If the standard-developing process meets criteria developed and published by ANSI, the standard can be accepted as an American National Standard. Of course, only one American National Standard can be accepted on each subject.

The standards process is voluntary in the United States. Whether a standard has a direct legal status completely depends on whether or not the standard has been adopted by statute within an established jurisdiction. An employer, a governmental body, or other organization may choose to embrace and implement any particular code or standard.

Like many other organizations involved with the production of standards, ANSI is a voluntary not-for-profit organization. It is not directly associated with government. However, the National Institute of Standards and Technology (NIST) has identified ANSI as the official U.S. representative to international standards efforts.

The ANSI objective, then, is to ensure a complete and open process for organizations that produce standards. ANSI ensures that rules associated with due process have been followed. In other words, every interested party has the opportunity to have his or her opinion considered in the entire development process.

■ THE *NATIONAL ELECTRICAL CODE*

Instead of a set of code or standards, as in OSHA, *NEC* is a single code. Technical committee members from industry volunteer their time and efforts to write this standard and keep it current. The National Fire Protection Association (NFPA), a nonprofit organization, is the publisher of the *NEC*. Because the *NEC* is a code, it has a scope rather than an objective. Like OSHA, the NFPA has an objective. Also like OSHA, NFPA codes and standards are tools used to accomplish the objective or mission of the organization.

The NFPA was organized in 1896 to improve the quality of life by reducing fires. The association assumed responsibility for a standard covering the installation of sprinkler systems. The mission grew and changed over the years as needs for standardization changed. The current mission of the NFPA is to reduce the burden of fire and other hazards on the quality of life by providing and advocating scientifically based consensus codes and standards, research, and education. The initial intent to improve the quality of life remains the primary goal today, but it has been expanded to electrical safety and other safety-related goals.

The *NEC* is a document that covers installation. The essential focus of the *NEC* is to define how equipment should be installed. Eliminating fires by identifying installation requirements remains a large part of the *NEC* content. Any installed service or equipment that meets the requirements identified in the *NEC* will be safe while the equipment is operating normally, provided it is adequately maintained.

As published by the NFPA and accepted by ANSI (as ANSI C1), the *NEC* is only advisory. The NFPA is a developer and publisher of standards. However, the *NEC* is written in language that allows it to be readily inspected and enforced. The *NEC* becomes mandatory when a governmental body adopts it. The enforcing authority is authorized by, and defined in, the law or ordinance that embraces the *NEC*.

■ NFPA 70E

The prime objective of this standard is to provide electrical safety-related work practices. Some industrial facilities might have installations that do not lend themselves to the same safety-related work practices that are appropriate in other locations. For instance, if a facility contains an electrochemical process, the process itself might act as a source of energy in such a way that the facility cannot logically be de-energized. Equipment grounding might be inappropriate, and even dangerous, in some conditions. NFPA 70E, *Standard for Electrical Safety Requirements for Employee Workplaces,* addresses these conditions.

This standard defines requirements that are based on the work task and exposure to electrical hazards, whereas OSHA regulations 29 CFR 1910, Subpart S, and 29 CFR 1926, Subpart K, define requirements based on the type of work environment, such as general industry, construction, or utility. On the other hand, NFPA 70E defines requirements that are based on specific electrical hazards and how an employee might be exposed to the hazards. NFPA 70E expects that an injury from an electrical hazard is not dependent on an individual work environment but that the risk of injury is equal from one work environment to another.

The objective of NFPA 70E is very similar to the objective of OSHA 29 CFR 1910, Subpart S, and 29 CFR 1926, Subpart K, with the exception that NFPA 70E is a *consensus* standard, and Subpart S and Subpart K are not. NFPA 70E is written and published through the same consensus process as each of the other NFPA standards. That process ensures equal participation by all interested sectors of the community. Adoption as an American National Standard certifies that the process is open and accessible.

Of course, change or review of either a new or revised OSHA standard, such as 29 CFR 1910, Subpart S, must follow the rules detailed in Section 6 of the OSH Act. On the other hand, the NFPA process for review and change is much more accessible to the public. NFPA 70E can be reviewed and revised as necessary to retain its relevancy. The revision process ensures that NFPA 70E addresses all hazards and work practices currently generally accepted by the community. For example, in Subpart S, OSHA contains neither information nor requirements for protection from either arc flash or arc blast. On the other hand, NFPA 70E addresses these hazards and offers suggestions about how to afford protection for people. NFPA 70E is considerably more protective from electrical hazards than OSHA standards.

■ THE *NATIONAL ELECTRICAL SAFETY CODE*

The IEEE publishes the *National Electrical Safety Code (NESC)*. Although substantially different from the NFPA standards process, the IEEE process is accredited by ANSI (as ANSI C2). The process ensures that affected interests are represented when a standard is generated or reviewed.

The objective of the *NESC* is similar to that of the *NEC,* but the scope of the document is quite different. The *NESC* addresses the utility industry and provides information intended to protect the general public from hazards associated with the transmission and distribution of electricity.

Transmission lines provide services to large segments of the public, including medical and other emergency facilities. Maintaining electrical service for emergency facilities is, of course, very important. Industrial facilities frequently require uninterrupted electrical service by contract. Work rules, as contained in the *NESC,* are intended to afford protection for utility workers as they go about their tasks of maintaining systems while the lines are energized, as is frequently necessary to maintain uninterrupted service. The attempt to provide uninterrupted business service effectively defines the working environment for many utilities.

Employers must avoid a working environment that expects and encourages employees to accept exposure to electrical hazards. The authors are not suggesting a weakness in the *NESC.* Instead, the process by which an employer adopts and applies the standard is very important. Because the *NESC* defines methods used to interact with energized transmission and distribution lines, most employers simply adopt these work methods as a "norm."

Employees should not be expected to accept unnecessary exposure to hazards. The requirements of all consensus standards, when adopted, should be implemented with full knowledge of how people are exposed to hazards. The working environment should consist of processes and work methods that minimize or avoid exposure to hazards.

■ INTERNAL CORPORATE STANDARDS

In some instances, large corporations establish and operate standards that apply internally only to the corporation. Internal programs usually have two objectives: control costs by defining specific engineering choices, and control exposure to safety, health, and environmental hazards. The documents produced by these efforts are sometimes called standards. They might also be called guidelines or procedures. However, the internal document usually serves in the same capacity as a standard.

Corporate legal organizations frequently are concerned about the name assigned to internal standardization efforts. Their concern is that if documents are called "standards," a disinterested party might expect all requirements of the documents to be implemented in all instances and in all locations. The legal concern is the potential use of the internal standards in litigation.

In the electrical discipline, programs that produce internal standards tend to select appropriate national consensus standards and offer interpretations of those requirements. If the external standard allows for a choice, the internal standard might make the selection for the corporation. Internal programs normally result in a high-quality installation and an excellent maintenance program.

DETERMINING EMPLOYER POLICIES

Every organization exists for a reason. Industrial and commercial organizations generally exist to make money. However, these organizations are made up of people who maintain values and beliefs. Safety policies usually are based on some element of the employer's values and beliefs. In this instance, employer values and beliefs refer to the composite management structure.

In many instances, employers' policies are driven solely by legal concerns. Establishing a sound defensive legal posture is important; however, effective safety policies are more important. Some employers believe that personal safety needs override the desire to establish a strong defensive posture. In these cases, the employer does not ignore legal concerns; instead, the company expends the greatest effort and amount of money on preventing incidents and injuries.

If a strong legal defense is determined to be the highest priority of a corporate policy, the standards policy should key on legally mandated regulations. Actions and discussions should reinforce this policy. On the other hand, if preventing injury is considered to be the highest priority, the policy should identify consensus standards (both legally mandated and not mandated) that the employer considers best able to provide adequate guidance to this end.

Again, real implementation of the standards policy involves the entire organization. It is important to note that legally mandated standards uphold the objective of preventing injury. Governmental units exist only to serve the public. When these units mandate a standard, such as the *NEC,* their intent is promoting public safety. In many instances, the intent of legally mandated standards is to prevent injury to members of the general public. In some organizations, such as OSHA, the intent is to prevent injury to industrial employees. The scopes of consensus standards must be reviewed to understand the mission.

Current editions of the following standards are recommended for inclusion in an employer's electrical standards policy:

- NFPA 70 (*NEC*)
- NFPA 70B
- NFPA 70E
- NESC
- OSHA 29 CFR 1910.147 of Subpart J
- OSHA 29 CFR 1910, Subpart S
- OSHA 29 CFR 1910.269 of Subpart R
- OSHA 29 CFR 1926, Subpart K

Employers with international facilities must consider international standards as they review SDOs for applicable products. Many countries have standards programs that are mandated by governmental action. Employers with international facilities must determine which standards apply to each facility. The standards policy included in the electrical safety program for each site should identify which international standards apply on that site.

SCANNING SDO LISTS FOR APPLICABLE PRODUCTS

By scanning the list of documents published by SDOs, a company representative easily can identify a few standards that cover any issue being considered. After identifying a particular standard of interest, the representative can scan each standard.

In some instances, several standards might address a single issue from different points of view. In this case, the SDO usually tries to ensure that the content of each standard is correlated with the content of each of the other existing associated standards. For example, in the NFPA process, several standards cover electrical issues:

- NFPA 70 covers electrical installation requirements.
- NFPA 70B covers electrical equipment maintenance.
- NFPA 70E covers electrical safety-related work practices.

It is important that these standards do not contradict one another. To have the best chance of avoiding contradictions, each panel or committee that writes these standards reports to NFPA through the Technical Correlating Committee (TCC). The TCC is charged with reviewing draft products of each panel or committee with an eye toward eliminating any contradiction. The TCC cannot correct any identified problem. Instead, the originating panel or committee must resolve identified problems.

REFERENCES

National Electrical Code® (ANSI/NFPA 70). Quincy, MA: National Fire Protection Association, 2002.

National Electrical Safety Code (ANSI/IEEE C2). New York: Institute of Electrical and Electronics Engineers, 2002.

NFPA 70B, *Recommended Practice for Electrical Equipment Maintenance.* Quincy, MA: National Fire Protection Association, 1998.

NFPA 70E, *Standard for Electrical Safety Requirements for Employee Workplaces.* Quincy, MA: National Fire Protection Association.

OSH Act: Public Law 91-596. Ninety-First Congress of the United States of America, S2193, Dec. 29, 1970.

OSHA Regulations 29 CFR 1910. Washington, DC: Occupational Safety and Health Administration, U.S. Department of Labor.

OSHA Regulations 29 CFR 1926. Washington, DC: Occupational Safety and Health Administration, U.S. Department of Labor.

Chapter 3

Setting Up the Electrical Safety Program

For most organizations, the initial program set up is difficult. Thinking about the entire program is somewhat overwhelming. However, any single piece of the program helps to get it started. One method of looking at the program is to envision the discrete elements as spokes or segments of a wheel (see Figure 3-1). A comprehensive program will consider the segments as discrete parts of the program. In order to keep the wheel rolling, all segments must be present.

It is possible for a single segment to exist, but all of the following segments must be present and adequately addressed for the electrical safety wheel to roll freely:

- Policies and procedures
- Site assessment
- Task assessment
- Personal protective equipment (PPE) requirements
- Hazardous boundaries and hazard/risk analysis
- Administration
- Lockout/tagout
- Training
- Auditing and recordkeeping
- Budgeting

This chapter provides an overview of these topics in the context of setting up the electrical safety program. Later chapters look at each segment of the program in detail.

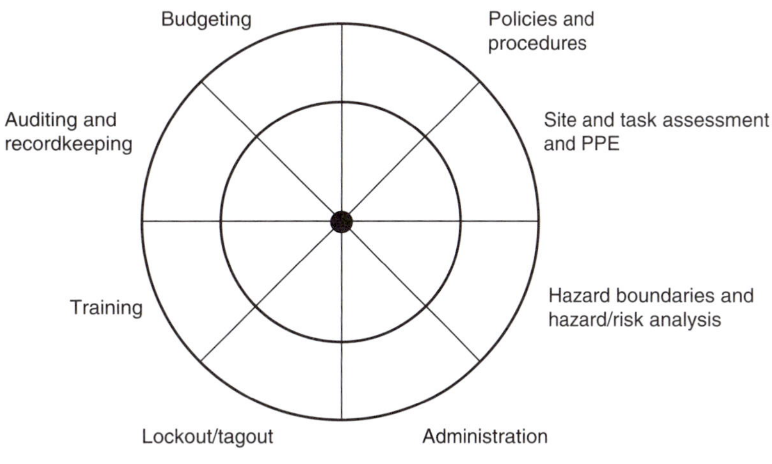

FIGURE 3-1 *The Electrical Safety Wheel*

■ PLANNING AN ELECTRICAL SAFETY PROGRAM

The electrical safety program will always be a work in progress. Many variables and outside influences will interfere with getting it started or with its forward progress. Most highly successful companies and programs have a history of numerous setbacks and, in some cases, situations where the entire company or project appeared to be destined for disaster. Overcoming negative issues with a positive approach and a dedicated focus on the program goals can achieve success. A critical component of planning a program is to justify the need for one in the minds of company management.

The Effects of Electrical Incidents on a Company

In the United States, 4,000 non-disabling injuries related to electrical shock require medical attention, and 3,600 disabling injuries occur annually (Department of Labor Statistics). In addition, each year 2,000 employees are sent to burn centers with severe electrical arc flash burns. Workers believe that they will come home safely every day from work, but 400 employees each year do not return home because they are killed in the workplace as a result of electric shock.

If an employee is injured, the severity of the injury can have a traumatic effect physically, emotionally, and financially. The majority of disabling injuries happen to employees aged 20–35 years. Although 3,600 disabling injuries from electric shock incidents are recorded yearly, no statistical data is available for electrical burn incidents. The industry uses the term "disabling injury." In reality, this term means that the victim of such an injury will never work in the same capacity again.

Many electrical burn injuries require extensive skin grafts, reconstructive surgery, or amputations. This concept, along with the knowledge that the majority of the employees injured are in the 20–35 year-old range, presents a picture of a devastated life, which could include social isolation through disfigurement, loss of self-worth, and in many cases, broken relationships and divorce. These workers also have lost their ability to earn a living forever.

How does an incident affect a company? Each company suffers far-reaching affects that must be considered. The first issue is dealing with the incident. The following questions must be answered:

- How does a company address an accident with employees and their families?
- Was negligence or inadequate training by the employer a part of the accident?
- Was a procedure in use for this task, and was it flawed or used incorrectly?

Addressing these issues is just the beginning of the company's problem. How does this incident affect the other employees? When OSHA becomes involved in an investigation of an incident, the burden of proof lies on the employer. Under OSHA rules, the employer, not the employee, can be cited. The employer is responsible for the action of the employee. This includes ensuring that the employee is aware of the hazards, is properly trained, and has the proper equipment to perform the task. The incident could also prompt a wider investigation into the company's safety program.

The company must consider more questions in its investigation:

- What are the hidden effects of an incident?
- How did this incident affect the other employees?
- If an employee is killed or injured, do other employees feel a loss of confidence in the company's safety program?
- How can the company assure the other employees that the procedures work properly?

Workers frequently are reluctant to perform specific tasks at the facility after a serious injury or fatality. A public relations problem might also exist. Municipal officials and local citizens are more likely to work cooperatively if a company is perceived as a good corporate neighbor.

If a corporation is considered environmentally unfriendly or perceived to operate without regard for the well-being of the town community, its operating costs could increase through fines, longer review cycles for permits, and associated municipal requirements. A local media report on a major incident can have a devastating impact on employees.

The financial costs of an injury to a company could include OSHA fines, medical costs, litigation costs, insurance increases, damaged equipment, lost production, wasted production, as well as time and resources to conduct investigations and implement remediation issues. While concerned employees are talking about the incident, they are not attending to their normal job responsibilities. If the incident results in a long-term injury or a death, the employee must be replaced, another factor that includes recruiting costs, training costs, and lost or diminished productivity.

Goals of the Program

H. W. Heinrich was a psychologist who changed the way the world considered safety fundamentals in the 1930s. He developed a theory that states for every 300 recordable injuries, approximately 30 lost-time injuries and one fatality will occur. Over the years, these relationships have proven to be relatively accurate. Some people feel that if the energy source is electrical, then a zero can be taken from the numbers, thus shrinking the triangle. This relationship might be used to help justify funding for an electrical safety program.

The Safety Triangle demonstrated by Heinrich's relationships in Figure 3-2 shows the relationship between behaviors and incidents. Developing an understanding of how these segments of the triangle relate to one another brings clarity to the strategies required for a comprehensive program. The triangle shows that, by not having a comprehensive safety program with a strong safety culture, the company will feel the negative effects of these behaviors in a matter of time.

If employees do not have adequate training for the electrical hazards involved, they are inviting disaster. If a company measures recordable injuries and establishes a program to deal with those injuries, they are, by definition, addressing lost-time injuries as well. The same relationship exists between incidents and recordable injuries. The next step—obviously—if a company program is based on monitoring and improving behaviors, is that incidents, recordable injuries, and lost-time injuries are all addressed.

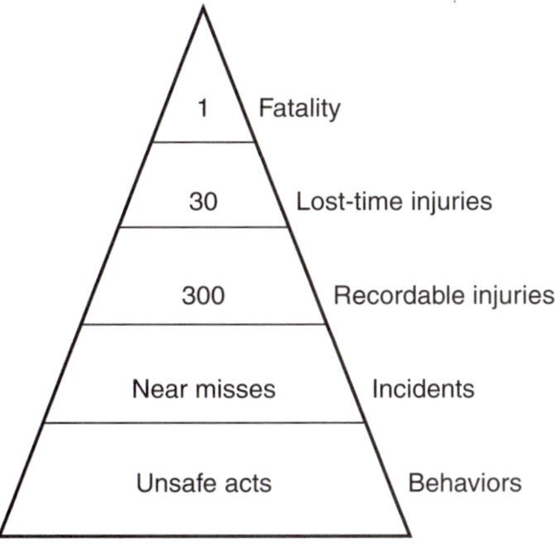

FIGURE 3-2 *Heinrich's Triangle Illustrating the Relationship of Behaviors to Injuries*

Heinrich's relationship suggests that a program that is focused on eliminating fatalities has a chance to be successful. However, if the program is focused on eliminating lost-time injuries, fatalities also will be eliminated. Likewise, if recordable injuries are eliminated, both lost-time injuries and fatalities will be eliminated. Efforts that are focused lower in the triangle have more data and stand a greater chance of success. Note that underpinning Heinrich's triangle are both near misses and unsafe acts. An effective electrical safety program will tend to place emphasis on unsafe acts.

The first priority of the planning process should be to generate an outline that defines the electrical safety goals for the organization. This outline should be consistent and align with the company philosophy on safety to enhance wide acceptance of the program. Electrical safety has obstacles that do not exist in other disciplines. These obstacles include hazards that are not readily visible, along with the stereotypes, misinformation, and lack of knowledge about electrical hazards that exist in the workplace. Many organizations rely on the electrical workers' training and experience as a substitute for safe work practices and proper protective equipment. Typically, a company would not let its employees handle harmful chemicals without proper procedures and protective equipment, but the same company might consistently expect workers to be exposed to energized electrical circuits without safe work practices and protective equipment.

Many organizations believe that all incidents and injuries can be prevented and have the goal of "zero incidents in the workplace." This is a realistic goal that can drive the processes of accident/incident reporting and accident investigation and analysis. Why should an incident in the workplace be an acceptable occurrence as part of a job task? Developing an electrical safety program with a proactive vision can result in lower operating costs and strong employee support. In addition to federal regulations, the company should incorporate into its program best work practices and techniques, both internally and within peer groups and organizations. Improving work practices is a critical process to ensure that the program is effective and up to date with new techniques and procedures.

The goals must be attainable and embraced by senior management for the program to have credibility. In addition, upper management must take an active support role for the program to be successful. The program plan must be attainable within the framework of the organization, and the mission statement should set the tone for the scope of the program and its components. The scope of the program is directly related to the budget and should spell out what steps, procedures, and policies will be required to attain the goal of zero incidents. The philosophy of management toward the electrical safety program should be spelled out in a procedure (see "The Electrical Safety Program" in Chapter 4, Procedures and Plans).

The Safety Culture

Every organization has a safety culture that needs to be taken into account in planning the electrical safety program. Employees and other stakeholders have an opinion about how much value the line organization has for personal safety. The program scope should be based on a realistic understanding of the existing culture. The program, then, should take steps as necessary to ensure that the safety culture is positive. Building a positive safety culture might require a long period of time. Top management has a major influence on the safety culture. As a matter of fact, in most cases, workers tend to watch the actions of managers instead of what they say. How managers act has the greatest impact on the safety culture.

Language and program principles should correlate with the scope and goals of the program. A good example would be to eliminate the word "accident" from the company vocabulary. When someone is injured or killed, an incident has occurred, not an accident. All incidents have a root cause and are preventable through proper techniques, training, and diligent effort. The risk of a worker being injured or killed while performing his or her duties simply must not be acceptable. Developing simple, concise concepts that are easily remembered is an excellent way to enhance the safety culture. Using concepts such as "Test before Touch" drives home a work practice that should be performed every time a person approaches a circuit component.

Two main factors should be considered when aligning the safety program with the desired safety culture. The first is ensuring that the mission or goal of the safety program is published in an internal document such as an employee handbook. In a smaller company, the publication is usually less formal, but more succinct. The second factor is the understanding that the culture is demonstrated through behaviors, actions, inactions, rituals, and values. Incorporating the electrical safety strategy into the existing strategy of the company's safety commitment, or as part of the company's culture to make it a "great place to work" environment, will contribute to buy-in from upper management and team building.

Packaging the technical aspects of the program with the proper culture creates a winwin situation with upper management. Generally, senior management has limited technical knowledge or understanding of the requirements of an electrical safety program. In many instances, a safety program is perceived as an added cost with virtually no economic benefit. Another frequent perception is that electrical safety has a negative impact on production. A good barometer of how to develop the scope is to understand how the company reviews similar items or programs in their budget. Most programs have a cost benefit component. Integrating the cost benefit into the justification for the electrical safety program is essential.

Cost Benefits

When planning the electrical safety program, two types of cost benefits must be considered: direct and indirect. Direct benefits are quantifiable, such as lower insurance cost through safe operations. Measurable benefits are essential to justify expenditure. The indirect benefits can be substantial for a company. Companies want to be perceived as a great place to work. A safe workplace has a far-reaching effect for a company. The concept of a good place to work helps with community relations, hiring and retaining a talented workforce, and municipality issues. Many companies are proactive in promoting good health for employees. Equating safety with good health under one umbrella delivers the message that the company cares about its employees. A well-trained employee performs his or her work with a higher level of confidence and increased productivity. The electrical safety program affects all of the employees, not just the people working on the systems.

■ SETTING UP THE PROGRAM

The first step in setting up an electrical safety program for any organization is to identify a person who will serve as the electrical safety manager (ESM) to build and maintain the electrical safety program. In some cases, the ESM might be a person already assigned this role by his or her job description.

The ESM will need skills that are comparable to those exhibited by a project manager of a large project. He or she should possess strong organizational skills, technical knowledge of the hazards involved in the work, effective interpersonal skills, and resourcefulness needed to accomplish the goals within the company structure. He or she should also have a passion for preventing injury from electrical energy sources.

The ESM should be forward-thinking and focused on the program goal. He or she should understand that setbacks might occur along the way and be prepared to resolve any problem that exists. The ESM must have the experience and knowledge necessary to develop goals and then design the program necessary to accomplish those goals. He or she must have sufficient knowledge and ability to generate management interest and buy-in to generate enthusiasm in the management structure.

The ESM must have the ability to overcome doubters who perceive an electrical safety program to be an unnecessary expense. He or she should be a participant in a network of experienced people to gather data to support the concept that an electrical safety program is a cost effective expenditure. The ESM will need to address claims that no real problem exists with the status quo. The ESM will need to overcome these and other negative attitudes in order to achieve a culture that embraces the electrical safety program.

Experience indicates that if the ESM can establish the cost benefits of the program, a manager is more likely to listen. A lost-time injury has a significant impact on insurance costs. It also has an impact on replacement costs. OSHA fines are another expense that impacts operating revenues. The ESM should use real data from the organization's history to establish normal costs of an incident and an injury. Using that data, the ESM should be in a position to sell the program to the line organization by comparing current incident costs with potential costs if the incident or injury rate is decreased.

The Role of the Electrical Safety Manager

The ESM should be responsible for establishing and maintaining the elements of the electrical safety program shown in the flowchart in Figure 3-3. The flowchart identifies separate functions that must be integrated into an overall program and also identifies where, in this book, information about each element can be found.

The ESM can use the flowchart to design and implement a new electrical safety program or to analyze an existing program. The existing program should be compared to each element illustrated in the flowchart. Any missing element(s) should be inserted into the existing program. Each element of the existing electrical safety program should be reviewed and compared to the information in the referenced chapter.

The ESM should ensure that each of the following program elements is adequately addressed:

1. Procedures and Plans
2. Site Assessment
3. Determination of Whether a Hazard Exists
4. Hazard Boundaries
5. Hazard/Risk Analysis
6. Personal Protective Equipment
7. Specialized Procedures (Including Lockout/Tagout)
8. Training
9. Budgeting
10. Program Administration
11. Auditing and Recordkeeping

Each of these program elements should relate to one another as depicted in the flowchart in Figure 3-3. A discussion of each element follows.

1. *Procedures and Plans.* Procedures and plans are an essential part of every electrical safety program. The program will need to have the procedures and plans that are essential

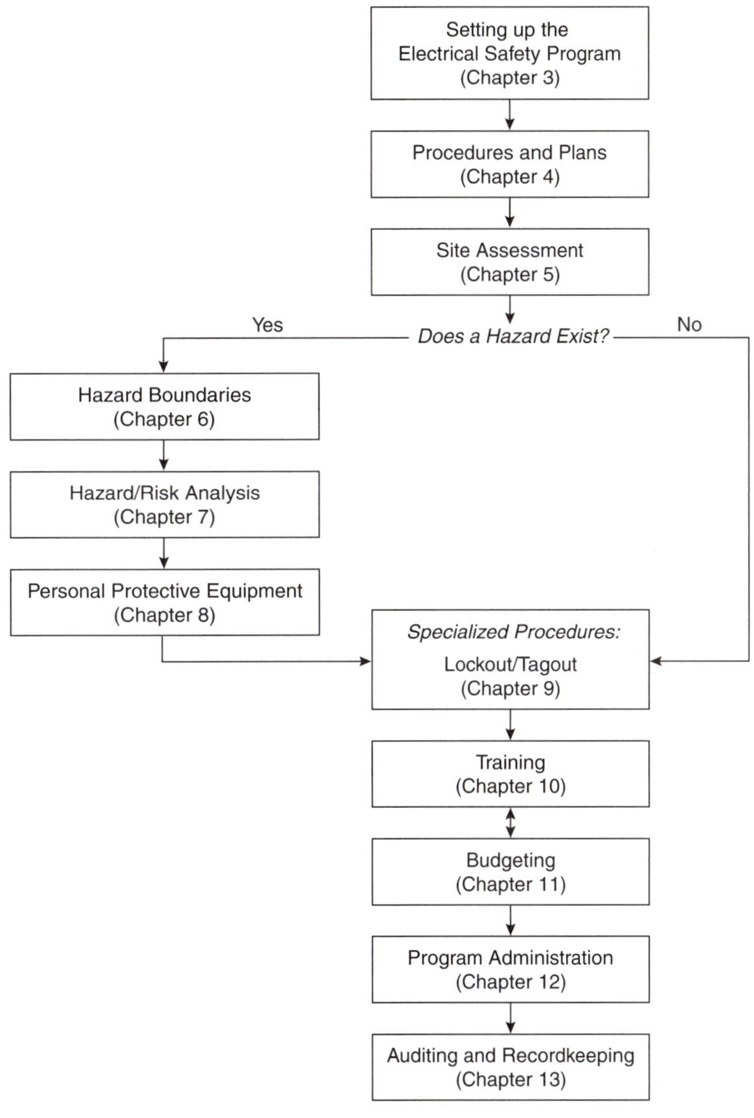

FIGURE 3-3 *Electrical Safety Program Flowchart*

to the safety of their workers and the safe operation of the equipment and processes. A procedure and plan should be defined and show how each one should be applied. There are potentially 40 procedures that may be required by a facility or company. See Chapter 4 for the definitions and background information required for procedures and plans. Chapter 4 also contains a list of the potential procedures with an explanation of each one.

2. *Site Assessment.* A site assessment is a key step in developing the electrical safety program. No two sites or companies are the same, hence the requirements and hazards would not be the same. The site assessment identifies the installation deficiencies, safe work practices, which procedures may be required, and tasks which might be done on or near live parts. Chapter 5 provides information to assist the reader to properly assess a site, protect workers from the hazards, and identify relevant topics for a training program.

3. *Does a Hazard Exist?* A site assessment must be completed to understand if a hazard exists. If a hazard exists, the program must follow the steps to determine the hazard boundary, perform a hazard/risk analysis, select proper personal protective equipment, and write a procedure or plan if applicable.

4. *Hazard Boundaries.* After the site assessment has been completed, the tasks that are hazardous in nature should be addressed. The first step in this process is to establish the hazard boundaries. Two different types of boundaries should be considered: shock boundaries to protect the worker from electrical shock and potential electrocution and flash boundaries that protect the worker from an arc flash incident. See Chapter 6 for more information on this subject.

5. *Hazard/Risk Analysis.* When a hazard/risk analysis is performed, all of the hazards associated with the task must be considered, and actions must be taken to minimize the exposure to the worker. Chapter 7 provides background information on the hazards and specific steps that can be taken to evaluate and analyze the hazard.

6. *Personal Protective Equipment.* PPE is the final barrier to protect workers from the effects of electricity. The effects and intensity of the hazard must be properly understood before PPE can be selected. The equipment must be adequate to protect the worker from the hazard and also comply with testing standards. Chapter 8 provides background information for the types of PPE that can be worn and also the standards that are used in the industry to ensure that the equipment is suitable for the purpose.

7. *Specialized Procedures (Including Lockout/Tagout).* After the site assessment has been completed, a list of potential procedures might need to be integrated into the electrical safety program. These procedures are critical in providing standardization in the manner in which all workers perform tasks. In addition to the identified procedure specific to the company or site, three procedures should be the staple of every electrical safety program: lockout/tagout, energized electrical work permit, and job briefing. See Chapters 4, 5, and 9 and Appendices C, D, E, and F for additional information on this subject.

8. *Training.* An employee training program is required to train all of the affected workers concerning electrical hazards in the workplace. This includes qualified and unqualified workers. The site assessment as outlined in Chapter 5 will identify tasks that are performed that could pose an electrical hazard. The program should develop a training module and provide training for the affected employees to address each of these situations.

9. *Budgeting.* Adequate funding is essential for the success of an electrical safety program. To develop a budget, two different areas should be explored: startup costs and maintenance costs. The startup costs, both direct and indirect, vary greatly from program to program. Realistically estimating the costs of developing this program is extremely important. Chapter 11 provides detailed information to help formulate and develop a budget, and Appendix K, Budget Samples and Checklists, provides a sample budget.

10. *Administration.* Administration is the implementation and maintenance of the electrical safety program. Chapter 12, Program Information Administration, provides the details and considerations for this part of a program, including the following:

- Updating the program book
- Maintaining reference drawings
- Training
- Auditing
- Recordkeeping

- Maintaining the budget
- Implementing the policies and procedures
- Documenting the program

11. *Auditing and Recordkeeping.* An audit is the most effective way to measure the effectiveness of a procedure. Without an audit, a company might not know that a potential defect exists in a procedure or plan until after an incident occurs. Recordkeeping is essential to understand the effectiveness of the electrical safety program, including procedures, plans, and training. Recordkeeping provides history that could be valuable in evaluating the qualifications of workers and also for litigation and regulatory issues. See Chapter 13 for more information on this subject.

Using the resources provided with this book can help any company plan for, provide the budget for, and administer a comprehensive electrical safety program.

■ REFERENCES

ASTM 1959, "Standard Test Method for Determining the Arc Thermal Performance Value of Materials for Clothing." Philadelphia: American Society for Testing and Materials, 1999.

National Electrical Code® (ANSI/NFPA 70). Quincy, MA: National Fire Protection Association, 2002.

NFPA 70E, *Standard for Electrical Safety Requirements for Employee Workplaces.* Quincy, MA: National Fire Protection Association, 2000.

U.S. Department of Labor Statistics. Washington, DC: Occupational Safety and Health Administration, U.S. Department of Labor.

Chapter 4

Procedures and Plans

The old sign Plan Ahead, in which the sign maker did not leave enough space for all of the letters, is comical, but it contains an element of truth. One of the guiding principles for any electrical safety program should be "Plan Every Job." The special considerations of work done on or near electrical circuits make the job-planning step a major investment in incident and injury prevention.

When a work task is not adequately planned, exposure to an electrical hazard, an unexpected event, or unexpected circumstances can result in an incident and injury. Sometimes an injury is not treated rapidly if the injured person is unable to summon help—another result of inadequate planning. Workers sometimes begin a work task by bypassing an interlock, opening a door, and jumping right into the task. This is an invitation for an incident to happen. All work should be planned *before* the work is started. In addition, when a work task reveals an unexpected condition, the work should be stopped and a new plan generated that considers the new information.

All electrical work should be guided by a plan. Written procedures and plans are similar documents. However, a worker might develop an effective plan without writing it down. The plan might be in the form of a standard practice that is used many times. The plan might be in the form of a written document that covers a single job or work task, or it might be a simple mental review of the work task. However, all electrical work should be planned. Experience shows that adequate planning prevents incidents and injuries. Procedures facilitate the planning process and ensure that corporate wisdom is considered as a plan is developed.

UNDERSTANDING AND APPLYING PLANS AND PROCEDURES

A prime objective of procedures and plans is to ensure that each worker completely understands the work task. As workers use words and phrases, a common understanding develops that enables workers to communicate easily with one another. However, sometimes understanding is limited to a relatively small community of people. For instance, one organization might use the word "practice" to mean the same thing that a different organization calls a "procedure." Sometimes, a common definition is restricted to a segment of a discipline. The term "hot work" is an example. Some organizations understand "hot work" to mean *electrically energized,* while other organizations might consider the phrase to mean a *thermal hazard.*

Note: *Although contact with an energized conductor can result in a burn, the temperature of the conductor is likely to be low.*

An organization, plant, or company should publish procedures, and the procedures should apply broadly within the unit that publishes it. Procedures should apply to all plant locations for a multi-site company and be available to every person on each site. Contractors should be contractually obligated to implement procedures that embrace the procedures in place on a site. If a contractor has an electrical safety program that contains procedures more restrictive than site procedures, contractor employees should implement the more restrictive procedures.

■ WHAT IS A PROCEDURE?

A procedure is a listing of steps that must be sequentially executed to accomplish a specific task or objective. The procedure documents corporate wisdom for future reference and application. A safety procedure should identify all steps necessary to accomplish the task mentioned in the scope. The procedure should specifically note each step that might require a worker to be exposed to a hazard and then identify any step that is necessary to control exposure to the hazard. If a different approach to the work can eliminate exposure, that option should be taken. If additional work steps are necessary to minimize the exposure, then the additional steps should be implemented. The procedure should then be modified as necessary to take advantage of the new information.

Every time a procedure is used to guide a work activity, the effectiveness and completeness of the procedure content should be reviewed. If a small change in the procedure could improve its effectiveness and/or understanding, the procedure should be changed and reissued. Ensuring this process results in a state-of-the-art procedure in which the entire organization is confident.

Sometimes procedures have objectives that are not directly related to safety. For instance, the prime reason to write a maintenance procedure might be to ensure that a minimum level of maintenance is performed on specific equipment or a specific class of equipment. Of necessity, maintenance procedures identify maintenance tasks. However, in many instances, a maintenance task will result in exposure of a worker to an electrical hazard. In this case, the maintenance procedure must either identify the work practices necessary to avoid the exposure or refer the maintenance worker to the appropriate safety procedure.

A procedure should be considered as a guide for a specific task, not as a routine, standard practice. A procedural task might be to operate a switch that has a hole in the door, whereas a standard practice would be to routinely stand to one side when operating a switch. The worker uses a published procedure as a guide to help him or her think about each step in the work task. Workers must recognize that an existing procedure might not directly cover the circumstances of the work task being planned. Therefore, a procedure should serve as a guide to accomplish the work task. The procedure could contain a step that is not necessary for the specific task being considered. However, if any step in the procedure is not implemented, a satisfactory explanation must exist that defines why a procedural step was not implemented. On the other hand, the work task might require a step that is not included in the procedure. This finding would constitute a reason to upgrade the procedure before it is used again.

■ WHAT IS A PLAN?

Like a procedure, a plan is a list of steps that must be accomplished sequentially for a work task to be complete. Whether a plan should be formal or not depends on the complexity of the work task. A simple task that only involves a single worker will probably not need to be in writing. It is not necessary for a work plan to be computer generated. A work plan that is written on the back of a discarded drawing may be effective.

The first requirement for an effective plan is that each step is indicated in the plan. The second requirement is for the plan to be critically reviewed by a second person who is knowledgeable about how to execute the work task. The third requirement is for all workers to be made aware of the plan, including if, when, and how workers might be exposed to an electrical hazard. Finally, the plan will be effective only if it is implemented as planned. However, if, when the plan is being executed an unexpected condition is found, the plan becomes null and void. A new plan must be generated that addresses the condition as it was found.

■ WHAT IS THE DIFFERENCE BETWEEN A PROCEDURE AND A PLAN?

Both procedures and plans are step-by-step instructions about how to execute a job. It is not necessary for every procedure to contain all information necessary to perform a task. A procedure usually provides general information, whereas a plan provides specific information. The plan might include using one or several procedures.

Many procedures begin their life as a work plan. When performing a repetitive job is necessary, the initial plan is modified to cover the new work. After the review and revision cycle occurs a few times, the plan takes on the characteristics of a procedure. The document can then be used in the future for similar work tasks.

It is not necessary for a work plan to be in writing. An effective plan can be mentally generated and verbally shared with other workers. One prime use for a work plan is to ensure that all workers associated with a job have the same understanding of how the work will proceed from its initial phase to its completion. If the plan can be shared and remembered by all personnel in the vicinity of the work task, then a verbally shared plan can be effective. In most cases, however, writing the plan on paper provides improved opportunity to achieve a common understanding among workers and supervisors.

■ WHAT ELEMENTS SHOULD A PROCEDURE CONTAIN?

A procedure must contain enough information for a user to be able to determine what is covered within the procedure and when the procedure should be used. For an organization, each procedure should have a consistent organization and format. Workers should be able to locate easily information that is needed. Procedures that have a consistent organization enhance their usability. A safety procedure should always include the following elements:

1. Scope or purpose of task
2. Qualifications and number of involved employees
3. Nature and degree of the hazards associated with the task
4. Limits or extent of the work task
5. Limits of approach
6. Safe work practices to be used
7. Personal protective equipment (PPE)
8. Insulating materials and tools
9. Special precautionary techniques
10. Electrical diagrams
11. Equipment details
12. Sketches/pictures of unique features
13. Reference or manufacturers' data

    1. *Scope or purpose of task.* The scope of a safety procedure should be relatively narrow. It is important that a user understand the requirements of a procedure in the context of a single objective. The scope statement should be direct and clear. Even though a procedure might have many discrete steps and require many pages to discuss all the necessary steps, the procedure scope should be restricted to a single objective. For instance, a procedure that covers creating an electrically safe working condition should not attempt to cover how to install lockout devices.

2. *Qualifications and number of involved employees.* In many instances, workers who implement a procedure should have unique knowledge about the work task. Perhaps the worker should have specialized or specific training related to the work task or be specifically authorized to perform the task. For example, workers who implement the lockout program are required to be specifically trained to understand the details of the program. The procedure should specifically identify necessary qualifications for workers who implement the procedure. Although workers should be trained in the use of procedures, they should understand that following a procedure can never take the place of appropriate training.

3. *Nature and degree of the associated hazards.* The basic purpose of any safety procedure is to decrease the chance of an injury. To accomplish that purpose, the procedure must identify each hazard to which a worker might be exposed while he or she is executing the work task. Once all of the hazards have been identified, the worker must understand the degree of hazard exposure at each step, as identified in the procedure. A procedure must make workers aware that the work task requires exposure to an electrical hazard. The worker must be aware of the exposure and how to avoid or minimize it. He or she must be able to select PPE that will afford the protection to avoid injury.

4. *Limits or extent of the work task.* Many documented injuries occur when a worker exhibits initiative to expand the scope of the work task. For example, in one incident a worker who was assigned to vacuum inside a compartment that is in an electrically safe work condition decided to vacuum the interior of an adjacent compartment. Although workers should be made aware of all circuits that are in an electrically safe work condition, the real need is for workers to know that *all other circuits* are energized.

5. *Limits of approach.* If any circuit or equipment is associated with other circuits or equipment that are not in an electrically safe work condition, the workers must be advised of those circuits. Both procedures and plans should identify all safe approach boundaries. Although workers should be trained to know if and when approach boundaries apply, the plan or procedure should contain all information related to any boundary that applies to the work task.

6. *Safe work practices to be used.* A work practice is a method that is normally used while executing a task. For instance, always standing to the right of a disconnect means when operating the switch handle is a work practice. How a worker selects a position for his or her body while performing a task is a very important aspect of satisfactorily completing a task. Performing a hazard analysis is a work practice. Although the ability to select and use safe work practices is an important aspect of training, each procedure should advise when a specific work practice is important. In the final analysis, however, the worker must be capable of selecting work practices that are appropriate for the type and degree of hazard associated with the task.

7. *Personal protective equipment (PPE) involved.* The primary purpose of a procedure is to provide each worker with advice about how to protect himself or herself from an incidental release of energy. While a procedure is being drafted, a preliminary hazard/risk analysis should be conducted. That hazard/risk analysis will identify specific PPE that should be worn when each step is performed. A worker might choose to wear additional PPE, but he or she must wear all PPE identified by the procedure.

8. *Insulating materials and tools.* If an electrically safe work condition does not exist, then at least one conductor remains energized. That conductor should be covered with insulating materials that are rated for the voltage. If the energized conductor cannot be effectively insulated from contact, the worker should wear rubber protective equipment that is rated for the voltage. Any tools that are required to execute the task should be insulated tools that also are rated for the voltage, even if the work does not involve intentional direct contact with the energized conductor. The procedure should identify both the protective equipment and the voltage rating for it.

Note: *"Rated for the voltage" means that the equipment, tool, or device has a rating that is established by a third party.*

9. *Special precautionary techniques.* The procedure should identify any special precautionary techniques that are necessary to protect workers from exposure to electrical hazards. For instance, if the overall job requires workers who are not associated with the work task to be in the area, then physical barriers and signs should be erected to warn them that electrical hazards are nearby. The work task might require emergency phones or other communication means to be installed.

10. *Electrical diagrams.* To ensure that workers involved in implementing the work task have all circuit information that is necessary to safely perform their task, the procedure should identify all diagrams or other drawings that depict the installation that exists. The procedure might not be able to identify specific drawings in the same way as a specific work plan; the procedure should identify drawings by type. The procedure should require that the drawings be current or be marked to depict the actual installation.

11. *Equipment details.* Sometimes equipment has a special latch, interlock, or other component. The procedure should caution workers to be advised that such components exist. The procedure should identify any mechanism that exists where details of this nature may be found.

12. *Sketches/pictures of unique features.* When an installation or system is complicated, the procedure should require that sketches or photographs be supplied to the workers before the work task is begun. Even though a qualified worker must be familiar with the construction details, he or she should be cautioned about any unique feature. Workers tend to rely on previous experience to anticipate construction details. If his or her recent experience does not include the type of equipment involved in the work task, the procedure should "jog" his or her memory about what to expect as the work task is performed.

13. *Reference or manufacturers' data.* Some electrical equipment is designed for a specific purpose and might contain normal equipment that is installed in a special way. For instance, if equipment contains a transfer switch, one or more switches might be energized from the bottom instead of from the top, as most workers would expect. Any reference information or data, such as set points provided by the equipment manufacturer, should be identified in the procedure. The procedure should specify where such information might be found.

▇ WHAT IS THE BEST PROCESS TO USE WHEN WRITING A PROCEDURE?

Effective procedures frequently begin as written plans for a specific work task. Executing a work plan always provides an opportunity to learn more about how the task should be performed. All work plans (and procedures) should represent the composite thinking of knowledgeable people who have experience with work tasks similar to the one being considered.

A single person should be assigned the task of producing an initial draft of an individual procedure. Sometimes, an outside consultant is engaged to provide an initial draft of a procedure. However, several people should be involved in the process of reviewing and generating a procedure. The following people should be involved in the process:

- Qualified person
- First-line supervisor
- Manager
- Second qualified person
- Safety professional
- Person knowledgeable of consensus standards and legal requirements

In some instances, a company might choose to charter an electrical safety committee and assign that committee with the responsibility of generating safety procedures for the company. Experience has proven this process to be highly effective. Electrical safety committees that comprise people who have experience in the roles identified above generally accept the challenge and take pride in the product.

In some instances, one or more procedures might be available from external sources. For instance, an appendix in NFPA 70E, *Standard for Electrical Safety Requirements for Employee Workplaces,* contains an example lockout/tagout procedure. An externally available procedure should not be implemented until each detail has been evaluated by the electrical safety committee, if one is chartered, or by people who are assigned the roles identified above.

Legally mandated standards assign employers the responsibility of providing procedures that are implemented by workers. Although the employer has this responsibility, only collective thinking by a group of experienced people can produce effective safety procedures.

■ HOW DOES THE SIZE OF THE ORGANIZATION IMPACT PROCEDURES?

The need for a procedure is completely independent of the size of the organization. However, large organizations should have procedures that are more formal. As the number of employees increases, it becomes more difficult to communicate with and among workers. It is possible for small organizations to define and communicate all procedure needs to all employees. Small organizations generally are provided more direct contact with senior management. However, even organizations with few employees need some written procedures. Both legal and consensus standards establish requirements for some written procedures. All organizations should provide and document those procedures.

As the number of employees and the complexity of the organization increase, the need for written procedures also increases. Procedures should be the place where corporate experience and wisdom are documented. If the procedures are adequately updated or otherwise maintained, the content of the procedures will indeed become the historical record of practices that have been successful for a specific organization.

Since procedures are the historical record of successful past work practices, procedures provide a very strong source of information that could serve as the base for both skill and knowledge training.

■ WHAT PROCEDURES ARE RECOMMENDED?

The following list contains procedures that the line organization should consider when they are generating an electrical safety program. The organization, plant, or corporation probably will not require every procedure mentioned in this list. A specific facility might also need a procedure that is not identified in this list. The organization should generate its own set of procedures that addresses the purpose behind these recommended procedures. This list can be used as a starting point. Following the list, each of these procedures is discussed in more detail.

1. The Electrical Safety Program
2. Electrical Diagrams—Updating and Using
3. Creating an Electrically Safe Work Condition
4. De-energizing Electrical Equipment
5. Developing Work Plans
6. Lockout/Tagout
7. Personal Protective Grounds
8. Selecting and Using Voltmeters
9. Purchase and Care of Personal Protective Equipment (PPE)
10. Labeling, Marking, and Identification
11. Switching Electrical Circuits (Normal and Emergency)
12. Electrical Test Equipment and Special Tools
13. Energized Electrical Work Permits
14. Conducting a Hazard/Risk Analysis
15. Conducting a Flash-Hazard Analysis
16. Work on or near Energized Electrical Equipment
17. Work on or near Energized Conductors and Circuit Parts
18. Inserting and Removing Units from Energized Motor Control Centers and Similar Equipment
19. Work on Energized Medium-Voltage Motor Control Centers
20. Ground-Fault Circuit Interrupters
21. Temporary Wiring
22. Electric Welding Machines and Portable Generators
23. Cable Tray Work
24. Stationary Battery Installations
25. Dismantling and Rearranging
26. Testing and Inspecting Electrical Equipment and Cables
27. Work on Energized 120/240-Volt Equipment
28. Work on Ungrounded Electrical Circuits
29. Work on Large-Capacity DC Equipment
30. Work on Variable Frequency Equipment
31. Portable Electrical Equipment
32. Using Mobile Equipment near Overhead Conductors
33. Safe Approach Distances
34. Training and Qualification
35. Electrical Qualification of Contractors
36. Underground Cables
37. Electrical Checkout and Startup
38. Electrical Incident Response
39. Housekeeping, Cleaning, and Storage
40. Auditing and Recordkeeping

1. *The Electrical Safety Program.* This procedure should document the overall electrical safety program. The procedure should clearly state all policies that senior corporate management expects to be implemented in all instances. If there are policies relating to documentation, standards, abandoned lines, working on energized conductors, or any other management expectation, the procedure should identify them, as well as corporate beliefs, principles, and philosophies that form the basis of the overall electrical safety program. This procedure should assign responsibility for generating and maintaining electrical safety procedures. In addition, this procedure should contain a set of common definitions used throughout the corporate electrical safety program. See Appendix A for additional information on outline considerations for an electrical safety program.

2. *Electrical Diagrams—Updating and Using.* This procedure should define how diagrams and other drawings are to be kept current. The procedure should assign responsibility to ensure that corporate intent related to drawings is performed. The procedure should identify where the most up-to-date drawings are located and how to retrieve the latest edition.

3. *Creating an Electrically Safe Work Condition.* This procedure should define the steps that are necessary to create an electrically safe work condition. The procedure should clearly assign responsibility for ensuring that an electrically safe work condition is the first alternative considered for all work tasks in the vicinity of electrical conductors and equipment.

4. *De-energizing Electrical Equipment.* This procedure should define the degree of planning that must exist before any switching is performed. If switches must be operated in a unique sequence, the procedure should identify those instances and clearly define the required switching sequence. The procedure should indicate a method to ensure that a particular sequence is followed.

5. *Developing Work Plans.* This procedure should provide guidance about how to generate an effective plan. The procedure should assign responsibility for generating a work plan and define how the plan will be reviewed to ensure that appropriate communication channels exist and are effective. The procedure should define that the plan includes steps to be performed in case of an emergency. The procedure should define the content of a work plan that is acceptable to the organization.

6. *Lockout/Tagout.* This procedure should define the steps that are necessary to implement an effective lockout or tagout. The procedure should define what constitutes an acceptable lockout or tagout and all expected requirements. It should include all mandated requirements and all requirements of national consensus standards. See Chapter 9 and Appendix F for additional information on Lockout/Tagout.

7. *Personal Protective Grounds.* This procedure should define acceptable parameters for ground sets that will be installed to operate overcurrent devices should the circuit be incidentally re-energized. The procedure should include ratings for protective ground sets. It should also contain a requirement for inspection and maintenance of safety ground sets.

8. *Selecting and Using Voltmeters.* This procedure should define safety requirements for voltmeters and list acceptable devices. It should define how to verify that the voltmeter is functioning properly. It should also define the criteria, such as consensus standards, that must be used to purchase any voltmeter for use on a site or in the organization.

9. *Purchase and Care of Personal Protective Equipment (PPE).* The procedure should specify acceptable PPE that will be purchased and made available to employees, including make and model. The procedure should define the process necessary for an employee to retrieve any PPE needed for any task on the site. It should also define when the equipment is to be inspected and how the inspection should be conducted. The procedure should explain all purchase specifications that embody appropriate consensus standards. See Chapter 8 for additional information regarding PPE.

10. *Labeling, Marking, and Identification.* This procedure should define how equipment and circuits are to be identified. It should provide specific guidance about how to assign identification and how the identification should be interpreted. The procedure should specify that labels and marking containing special instructions are to be constructed and where such labels are to be placed. It should also assign responsibility for maintaining labels and marking in a legible condition.

11. *Switching Electrical Circuits (Normal and Emergency).* This procedure should provide guidance about who has the authority to operate electrical switches. It should establish ownership of the electrical equipment. The procedure should define normal operation and emergency operation.

12. *Electrical Test Equipment and Special Tools.* This procedure should define acceptable criteria for electrical test equipment and how such equipment is to be maintained and inspected. The procedure should identify acceptable test equipment by make and model. If third party testing is required, the procedure should identify which test laboratories are considered to be acceptable. If special tools are required for any work task, the procedures should identify those tools and specify how the integrity of the tools is verified.

13. *Energized Electrical Work Permits.* This procedure should identify when authorization is necessary to perform a work task that is considered to be electrically hazardous. The procedure should establish a protocol to be implemented when any electrical work task must be conducted in the vicinity of energized electrical conductors or circuit parts. It should define a requirement for a written plan to be drafted before executing an energized electrical work permit. The procedure should identify the person (either by name or title) that has the responsibility to authorize work that has an elevated risk of injury, identify when such authorization is required, and identify the form and format of the authorization. The procedure should also require the signature of the responsible person and concurrence or agreement by each worker that the plan is acceptable. See Chapters 5, 6 and 7 and Appendix E for additional information regarding energized electrical work permits.

14. *Conducting a Hazard/Risk Analysis.* This procedure should define the process of conducting a hazard/risk analysis and assign the responsibility of ensuring that the hazard/risk analysis is performed by a specific person. See Chapter 7 for additional information regarding Hazard/Risk Analysis.

15. *Conducting a Flash-Hazard Analysis.* This procedure should define the acceptable process of conducting a flash-hazard analysis. If a computer program is to be used to conduct the analysis, the procedure should identify the computer program. If manual calculations are to be used, the procedure should contain the equations that are to be used in conducting the analysis. See Chapter 7 for additional information regarding Hazard/Risk Analysis.

16. *Work on or near Energized Electrical Equipment.* This procedure should define the conditions and type of work tasks that may be performed on or near electrical equipment that is either still in service or otherwise energized.

17. *Work on or near Energized Conductors and Circuit Parts.* This procedure should define the circumstances that must exist before this type of work task is performed. The procedure should define a requirement for a written plan and a requirement for the work to be authorized by senior management. The procedure should define the qualifications of all personnel that will be involved in the work. It should also contain a requirement to stop work and generate a new work plan if unexpected conditions are observed or otherwise identified.

18. *Inserting and Removing Units from Energized Motor Control Centers and Similar Equipment.* This procedure should define requirements necessary before a worker is allowed to insert or remove a single unit (plug-in-style) into or out of a motor control center that has energized bus. The procedure should define how to inspect the unit and perform tests that reduce the risk of a short circuit while withdrawing or inserting the unit. If authorization and/or PPE is required, the procedure should so state.

19. *Work on Energized Medium-Voltage Motor Control Centers.* This procedure should identify precautions and define requirements that must be taken to avoid injury from electrical hazards. One primary purpose for the procedure should be to identify acceptable limits for work activity. The procedure should specify any specific personnel training requirements and specifically discuss any possible unique hazard exposure, such as current transformers.

20. *Ground-Fault Circuit Interrupters.* This procedure should define requirements that establish how and when ground-fault circuit interrupters (GFCI) should be used. The procedure should also establish criteria that define acceptable third-party testing requirements. If locations exist where GFCIs should not be used, this procedure should identify those locations and then define the alternative protective schemes to be used.

21. *Temporary Wiring.* This procedure should define authorizations (if any) to install a temporary circuit. It should establish the method to control the duration of the temporary circuit and define how the temporary conductors are to be identified. If the temporary wiring introduces the potential to backfeed existing equipment, the procedure should define requirements to alert workers that potential exposure exists.

22. *Electric Welding Machines and Portable Generators.* This procedure should define any unique requirements for electric welding machines and portable generators and how the welding circuit must be grounded. The procedure should consider a requirement to install both a supply and return conductor for the welding circuit. It should establish any special requirements for high-frequency welding or portable generators. If special grounding is necessary, the procedure should define that grounding.

23. *Cable Tray Work.* This procedure should define how to safely add or remove one or more cables when the work task involves a cable tray that is in service. The procedure should define how to determine if an electrical hazard exists and how to mitigate potential exposure to the hazard. It should also define any necessary protection when the work task is nearby or over an existing cable tray. If the voltage or age of cables in the tray has a bearing on protective requirements, the procedure should define what actions are necessary.

24. *Stationary Battery Installations.* This procedure should identify any unique work method associated with a battery bank and explain how to replace an individual cell. The procedure should also address any chemical hazard that might be present and discuss protective equipment that is required, if any.

25. *Dismantling and Rearranging.* Dismantling an existing facility could expose unqualified persons to electrical hazards. This procedure should establish a process to determine whether or not a building or equipment to be dismantled is in an electrically safe work condition. If specific qualified persons are required for the work, the procedure should either name the individuals or define their qualifications.

26. *Testing and Inspecting Electrical Equipment and Cables.* This procedure should address the special hazards associated with these activities. If special experts are required to implement the inspection or testing or interpret the findings, the requirements should be specified in this procedure. The procedure should indicate when and how special communication systems are necessary.

27. *Work on Energized 120/240-Volt Equipment.* This procedure may provide special authorization to perform work on or near equipment in this voltage range. The procedure should indicate any special precautionary measures that must be taken and specify what PPE is to be used while performing the tasks covered by the scope of the procedure. If specifically trained personnel are required, the procedure should identify the specific training requirements.

28. *Work on Ungrounded Electrical Circuits.* This procedure should address the unique methods of potential exposure to electrical hazards that accompany ungrounded circuits and clearly indicate where such circuits exist. It should also suggest how usual safe work practices might need to be modified for use when the electrical circuit is ungrounded. If specifically trained personnel are required to perform the task, the procedure should identify the necessary training.

29. *Work on Large-Capacity DC Equipment.* This procedure should address the special hazards associated with high-capacity direct-current circuits. The procedure should require that workers have special training that covers both the unique hazards and any unique exposure that the DC system encompasses. If the high-current system is related to electrochemical processes, the procedure should be specific about acceptable work practices and about conductive equipment.

30. *Work on Variable Frequency Equipment.* This procedure should provide guidance for employees who are assigned tasks that involve variable frequency equipment. Hazards and potential injuries change significantly as the circuit frequency increases or decreases, and this procedure should address these variations of hazards. The procedure should provide for specific training that generates understanding of the hazards associated with variable frequency equipment.

31. *Portable Electrical Equipment.* This procedure should define all inspections and tests that are required to be performed on electrically powered portable hand tools. The procedure should specify any required purchase specification (such as double-insulated tools) that is required by the safety program. The procedure should specify recordkeeping requirements for inspections and tests.

32. *Using Mobile Equipment near Overhead Conductors.* This procedure should define requirements associated with using mobile equipment in the vicinity of both insulated and uninsulated overhead conductors. The procedure should define requirements for insulated and rated mobile equipment, such as line trucks, and for normal transportation equipment. The procedure should define what precautions are necessary if mobile equipment is required to stop in the vicinity of overhead conductors.

33. *Safe Approach Distances.* This procedure should specify and discuss approach boundaries that are to be used by a site or organization. The procedure should cover the differences between boundaries related to the shock hazard and boundaries associated with arcing-fault hazards. If more than one method of generating this information is acceptable, the procedure should cover each method. If approach distances are predetermined and equipment appropriately labeled, the procedure should describe the correlation between information on the labels and the method of generating the information. See Chapter 6 for additional information regarding Hazardous Boundaries.

34. *Training and Qualification.* This procedure should address training requirements for both qualified and unqualified persons and clearly indicate requirements for retraining. The procedure should define what training records are required and how long they should be retained. It should discuss the need for each person on site to understand the limitations of his or her own qualifications. The procedure should define how a supervisor or worker could interact with the training process. See Chapter 10 for additional information regarding Training and Qualifications.

35. *Electrical Qualification of Contractors.* This procedure should define requirements that are expected to direct outside contract employees, how safety expectations are communicated to the contractor, and how the contractor communicates concerns. It should define if, and how, contract employees are monitored for safety concerns. If formal documentation is considered necessary, the procedure should define the necessary documents. The procedure should also define who is authorized to act on behalf of the contractor and the owner.

36. *Underground Cables.* This procedure should define how to determine the location of underground cables. It should establish whether or not authorization is necessary for underground excavation. If special cable-locating equipment is necessary, the procedure should clearly specify the type and make of the equipment. If external consultation is required, the procedure should describe how to engage the consultant.

37. *Electrical Checkout and Startup.* This procedure should define a process to communicate "ownership" of electrical equipment. When new equipment is installed or existing equipment undergoes significant modification or maintenance, defining a method to determine which organization has control of the equipment or circuit is one critical element of preventing injury. This procedure should define that process.

38. *Electrical Incident Response.* This procedure should describe actions to be taken in the event of an electrical incident. It should describe the circumstances that require external emergency medical assistance and define information that should be provided to medical personnel.

39. *Housekeeping, Cleaning, and Storage.* This procedure should define responsibility for housekeeping in electrical equipment rooms. It should also define who is responsible for housekeeping where potential exposure to exposed energized electrical conductors or parts exists.

40. *Auditing and Recordkeeping.* This procedure should specify all requirements for audits of the electrical safety program. It should assign responsibility for ensuring that required audits are performed. The procedure should require specific corrective action for irregularities. The procedure should specify what records are necessary and how long such records are to be maintained. See Chapter 13 for additional information regarding Auditing and Recordkeeping.

■ REFERENCE

NFPA 70E, *Standard for Electrical Safety Requirements for Employee Workplaces.* Quincy, MA: National Fire Protection Association, 2000.

Chapter 5

Site Assessment

Electrical safety programs are like used cars. Many of them appear to be similar, but no two are exactly the same. A systematic and methodical approach is required to fully understand the information and logistics required to provide a comprehensive safety program that protects all workers. A site assessment is the most effective way to gain an understanding of the scope, magnitude, and requirements of the program.

One-third of all electrical incidents are caused by faulty equipment, and two-thirds of the incidents are caused by unsafe acts. Appendix B contains two site assessment checklists that can be used as a template to assist in compiling the required information to develop and implement a comprehensive program. Using these checklists enables the user to evaluate and consider the potential hazards that affect the workers and provide a proactive remediation of all of the issues.

The checklists in Appendix B are the General Industry (GI) Checklist and the Construction Checklist. The site checklists are each divided into two sections: installation and safe work practices. Several conditions affect a safe installation, but responsibility is a key factor over time. The workers are qualified and the installation is completed in accordance with all applicable codes, using safe work practices. After installation is complete, the authority having jurisdiction signs off on the initial installation. But then, who is responsible for the modifications, repair, and maintenance of the system? This work may not have been inspected for code compliance, resulting in modifications and repairs that might create unsafe situations.

■ THE GENERAL INDUSTRY CHECKLIST

The first part of the GI checklist addresses all of the physical equipment for a site. The user can perform a complete walkthrough to verify that all of the equipment is installed in accordance with the applicable codes and does not expose the worker to a hazard. The checklist is set up in logical order, starting with the service entrance equipment and proceeding throughout the entire electrical system. The second part of the checklist addresses safe work practices, which are addressed in a separate section later in this chapter.

Initial Inspection Activities

The Installations section of the GI checklist is subdivided into different types of equipment, starting with major equipment, including substations, switchboards, and panelboards, and continuing through the rest of the systems. One of the initial inspection activities is to determine whether the equipment is suitable for the location or the environment. For example, the National Electrical Manufacturers Association (NEMA) has classifications for the different occupancies. The equipment is classified as suitable for outdoor or indoor use. It has further subcategories for specific conditions such as rain, water, oil, and corrosive agents; for instance, NEMA 1 equipment is suitable for indoor dry locations, and Section 430.92 of the 2002 *National Electrical Code*® (*NEC*®) provides a table that correlates the NEMA classifications with the characteristics of specific environments.

Another inspection activity verifies whether the equipment has been tested and approved for the purpose. Section 110.3(A) of the *NEC* provides the criteria for suitability of the equipment. The approval criteria include mechanical strength and durability, wire-bending and connection space, electrical insulation, and classification by type, size, voltage, current capacity, and specific use. A label from a third-party inspecting agency, in many cases, will verify this suitability.

Live Parts

Since it is crucial to protect workers from exposed live (energized) parts and associated hazards, inspection activities must verify that all of the covers are on the equipment. Replacing covers in a timely manner after completing the work on equipment should be incorporated into the company procedures and culture. If the equipment does not have covers, the worker must be protected from contact with the equipment by a fence, locked door, or suitable barrier. Warning signs to alert workers of the potential hazard must also be in place.

Space Requirements

Inspection activities must also focus on the working space around equipment. The space requirements in Sections 110.26(A) and 110.34(A) of the *NEC* provide the minimum space requirements for equipment. The intent of these two sections is to provide adequate space for personnel to perform any required operation without jeopardizing worker safety. These equipment operations include the following:

- Examination
- Adjustment
- Servicing
- Maintenance

Examples of the equipment include the following:

- Panelboards
- Switches
- Circuit breakers
- Controllers
- Controls on heating and air-conditioning equipment

Included in these clearance requirements is the step-back distance from the face of the equipment. Table 110.26(A)(1) in the *NEC* provides requirements for clearances away from the equipment, based on the circuit voltage-to-ground and whether or not grounded or ungrounded objects exist in the step-back space or exposed live parts across from one another. The voltages-to-ground consist of two groups: 0 to 150 volts, and 151 to 600 volts, inclusive. Additional space could be considered to protect the worker from being trapped in an arc-flash incident. The typical arc-flash boundary for a 600-volt system is 4 feet.

Additional Installation Requirements

As users perform the site assessment, the checklist will navigate them through the site to evaluate all of the issues that address worker safety. The paragraphs below follow the general chronology of the checklist and provide background information to assist in the site assessment.

Dedicated space above electrical equipment is required to permit the installation or modification to the electrical circuits in a safe manner. Storing materials in front of electrical equipment impedes workers from safely operating the overcurrent protection devices in an emergency situation and also may be a potential fire hazard. Identifying the panels and branch circuits serves many purposes, such as assisting in troubleshooting equipment in a timely manner, saving valuable time in an emergency situation, and providing clarity for lockout/tagout procedures.

Closing unused openings also prevents two hazards. First, with the unused openings closed, the worker is not able to touch the live parts with his or her body or conductive objects, and second, having the holes closed prevents foreign materials from infiltrating the equipment. An example is a machine shop that has a metal grinding operation. The metal dust could infiltrate the equipment, creating a path-to-ground, which could cause a short circuit, arc flash, explosion, and potential injury to a worker. In addition, adequate illumination reduces the chances of committing an unsafe act.

Another installation activity is appropriate termination of conductors. If the conductors are terminated correctly, heat and potential fire hazards are reduced. Damaged conductors are always a potential hazard for electric shock and arc flash.

All of the wiring methods require similar investigative methodology to ensure safety. The wiring methods must be approved for the purpose, location, and environment. For example, flexible cord, such as extension cords used for appliances and portable tools, should be of the extra-hard usage type.

Wiring devices should be of the grounding type, according to Section 406.3(A) of the *NEC*. Section 210.8 defines the minimum requirements for ground-fault circuit interrupters. Do areas in the company exist where ground-fault circuit interrupters could provide additional safety for the workers? A receptacle located in the proximity of a sink on a production floor or janitor's closet is an example of such an area.

Generators and their associated equipment use the same review criteria as service equipment and panelboards. Some of the common issues are whether or not the equipment and its loads are protected for overcurrent, and whether or not the equipment has a disconnecting means.

Grounding is an important safety consideration. One of the main functions of the grounding system is to facilitate the proper operation of the overcurrent protection devices. If the grounding system is inadequate, the device can take longer to open, putting the workers and equipment at increased risk through electrical shock and arc-flash hazards. The grounding path must be permanent and continuous. The grounding conductor must be adequate to carry the ampacity of the fault until the overcurrent protective device can open the circuit.

If hazardous locations exist within the company, they require special consideration. Is the location properly classified? Articles 500 through 505 of the *NEC* cover the requirements for electrical installations classified as hazardous locations due to the materials handled, processed, or stored in those locations. Some of the most common materials encountered in hazardous (classified) locations are flammable liquids. A flammable liquid is one that has a flash point below 100°F. The equipment in these locations must be approved and be marked for the specific occupancy. In many cases, hazardous locations require more stringent wiring methods to ensure safety.

■ THE CONSTRUCTION CHECKLIST

Construction requirements are often the most overlooked installation items for a company. Construction sites could include the temporary lighting and power for a building under construction. It could also be the lighting and power requirements for remodeling, emergencies, and tests. The words *construction* and *temporary* imply that a lower level of quality is required than for a permanent installation. Before the 2002 edition of the *NEC*, the scope for temporary wiring stated that "this article applies to temporary electrical wiring and power methods that may be a class less than would be for a permanent installation." Although the installation may be temporary, the electrical hazards are the same as those of a permanent installation. Temporary installations are not necessarily installed by a construction contractor. At the same time, construction contractors are not installing only temporary facilities.

In some cases, exposure to the hazards creates even more dangerous conditions. Typically, the worker might encounter water, poor lighting, and other adverse conditions. Temporary wiring methods must be approved based on the following criteria:

- Length of time in service
- Severity of physical abuse
- Exposure to weather
- Other special requirements

Installation Follow-up

The construction checklist addresses physical equipment and safe work practices required for a site. Temporary installations require a complete site walkthrough and evaluation of procedures to provide safety for all workers. All equipment must be installed in accordance with the applicable codes. After the initial installation is complete, system maintenance is essential. Leaving covers off or using cardboard for a protective barrier for live parts are typical examples of poor maintenance techniques. Traditionally, the authority having jurisdiction (AHJ) inspects the initial temporary service and the branch wiring system and authorizes the system to be energized. The AHJ normally does not return to inspect the modifications to the system. Developing a work plan that provides adequate safeguards to protect the worker is required.

In many instances, the temporary installation is performed by an outside contractor. If the work is in an existing, occupied facility, all workers must be protected. Typically, this work is done with less planning, direction, and supervision as compared with work done in a permanent installation. All electrical installations, including temporary wiring, should be installed in accordance with all applicable codes, including the requirements of the *NEC*. Article 527 of the 2002 edition of the *NEC* permits modifications to the general rules of the *NEC* for temporary installations. Relaxing the general installation requirements changes the degree of exposure to electrical hazards. These requirements should be incorporated into a temporary installation working plan and reviewed by all of the contractors and their employees to provide uniformity for all installations.

The checklist follows the same criteria as the General Industry checklist. Even though the installation is classified as temporary, the same electrical hazards exist. A walkthrough verifies that all of the equipment is installed in accordance with the applicable codes and that the equipment does not expose the worker to a hazard. Several variables affect the safety and integrity of the electrical system.

Specific Requirements for Temporary Installations

Service equipment includes substations, switchboards, panelboards, and other major equipment. This equipment must be suitable for the environment and also the conditions caused by an expected lower level of housekeeping and maintenance.

Equipment and the overall installation must meet the same requirements as a permanent installation, as defined in Section 110.2 of the *NEC*. Section 110.3(A) provides the criteria for suitability of the equipment.

Cords and receptacles that are used as a temporary installation are required to provide a higher level of assurance that the integrity of the ground-fault return path is good and is maintained in a good condition. An assured equipment grounding program is identified in some consensus standards as an acceptable means of providing this assurance. Alternatively, continuous use of a ground-fault circuit interrupter for all instances where a portable cord or a portable tool is used provides the same level of assurance that the worker is protected from electrical shock. For details of an acceptable process, see OSHA 29 CFR 1926.404(b)(1)(iii).

■ SAFE WORK PRACTICES (FOR BOTH CHECKLISTS)

Safe work practices are dependent upon the hazard and the worker's exposure to the hazard. The exposure is dependent solely upon the work task. Safe work practices are not dependent upon the type of work, whether the installation is temporary, construction, general industry, or utility. For that reason, safe work practices have been added at the end of both the checklists in Appendix B. The information is provided in each document for the convenience of the user.

Because unsafe acts account for two-thirds of all electrical incidents, safe work practices should be the heart of the electrical safety program. The safe work practices are formulated through a four-step building-block process:

- Identify the hazard.
- Use procedures and plans.
- Establish a safety culture.
- Support the program with a budget.

When the facility identifies the hazards present on its site(s), it can begin to know what must be done to protect the workers. Then management can plan the necessary procedures and plans. An important factor for a safe workplace is to develop a safety culture that integrates the safe work practices into the daily routine of the company. Finally, a budget must be in place to support the program.

A written safety program is the most effective way to convey the policies, procedures, and working ethics of a company. A written program is a living document that evolves and is enhanced over time. Input from all of the affected parties is required to ensure safety for everyone.

Designating a person in charge brings many benefits to the program. The person becomes a focal point whereby information is gathered, compiled, and disseminated. The designation reinforces the company's commitment to provide an electrically safe work environment. It also provides continuity for the development, implementation, and maintenance of the program. One of the responsibilities of the designated person is to budget adequate funding to support this initiative.

A new car has a limited life and reduced reliability if it is not maintained. The same principle should be applied in maintaining the safety program book and associated records. A good working plan ensures that qualified persons perform electrical tasks, that safety meetings are conducted, and that records are maintained to analyze the proficiency of the program policies, plans, and procedures. Chapters 12 and 13 provide the information to assemble the electrical safety program book and associated records.

Do the contractors and vendors follow the same procedures and plans as the workers in a company? If they do not, contractors/vendors not only put themselves at risk, but the company's workers and financial stability as well. The electrical safety program must reinforce these concerns through written policies, training, and prequalification of contractors and vendors.

Working on or near Live Parts

The most obvious electrical hazard to workers is working on or near live parts. Many people oversimplify or minimize these hazards. Every task should be analyzed individually, to ensure that the worker is properly protected. Any task performed on exposed live parts at 50 volts or more puts the worker at risk. This is the same voltage level that the OSHA regulations use as the threshold of a potential electrical shock hazard. Electrical hazards should be given greater respect than other hazards in a company.

Training

Training is required to ensure that the qualified persons can perform the tasks safely and that unqualified persons are aware of the hazards and know how to avoid them. How can a company determine that the contractors and vendors are properly trained? A specification is required to make sure that the contractor's training is aligned with the hazards present and the remediation techniques of the company. In addition to electrical safety training, do other hazards exist in the company? Subject matter for training might include the following:

- OSHA injury and illness recordkeeping
- Walking/working surfaces
- Fire protection and other life-safety issues
- Hazardous materials
- Respiratory protection training
- Occupational noise and hearing conservation
- Permit-required confined space
- Machine guarding
- Material handling
- Incident investigation

All hazards must be identified and plans and procedures in place before a training program can be completed. Having all workers complete a 10-hour OSHA safety course for general industry provides a broad base of safety awareness for all workers. Another consideration is who should provide the safety training. Provision of the training, whether done on site by internal people, outside the company, or having a consultant provide the training depends on the resources and qualifications of the company and the budget. The frequency of the training varies widely depending on the qualifications of the personnel and their proficiency. The frequency of the task can vary from subject to subject and person to person. Maintaining training records is critical when an incident occurs or an OSHA inspection is performed. See Chapter 10 for additional information on training.

Procedures

Written procedures are essential to the safety program for both working and training purposes. The safe work practices checklist of the *General Industry Checklist* in Appendix B contains over forty potential procedures that could be used. In the site assessment, each of these procedures should be investigated to see if it is applicable to the site. The need for a written procedure depends on the qualifications of the workers, continuity of the workforce, and the company's culture. See Chapter 4 for additional information on Procedures and Plans.

Lockout/Tagout

NFPA 70E, *Standard for Electrical Safety Requirements for Employee Workplaces,* uses the term "electrically safe work condition" as the appropriate way to provide optimum worker safety. Lockout/tagout (LO/TO) is an integral part of this process. OSHA has specific requirements for LO/TO. Several fundamental questions must be answered in the LO/TO program, and steps must be taken to ensure workers' safety.

OSHA regulations for construction do not have the same LO/TO requirements as general industry, making the worker more vulnerable to injury. Using the procedures in the general industry standard (29 CFR 1910) ensures a higher level of safety. The general industry standard provides clarity to the requirements. A written procedure is always required and must be readily available for all affected workers. The LO/TO procedure must address all of the energy types in the company. The procedure must be audited annually and records kept and maintained. In addition, the affected workers must have the proper LO/TO equipment and personal protective equipment (PPE) to perform the procedure. The LO/TO equipment should be readily available and uniquely distinguishable to workers. See Chapter 9 for additional information on Lockout/Tagout procedures and Appendix F for a sample Lockout/Tagout procedure.

The Energized Electrical Work Permit

One of the procedures is an energized electrical work permit, a copy of which is included as Appendix E. This permit is invaluable to the company. First, it establishes the necessity to perform this work while the circuit remains energized. The worker and a responsible person (perhaps the equipment owner or another member of senior management) must sign this form. Once the necessity is established that the work must be performed while the circuit is energized, the worker is assured that due consideration has been given to the situation. The responsible person has indicated, by his or her signature, that the company has acknowledged the increased risk and accepted the increased cost associated with working on the energized circuit (existing location).

The permit requires that a detailed analysis be performed to establish the electric shock and arc-flash exposure. Appropriate PPE must be selected based on the result of the detailed analysis. The permit should identify how unqualified persons should be warned to stay clear of the work area. A job briefing should familiarize the workers with the task and the boundaries of the task and any specific issues that are unique to the work. And, perhaps most importantly, it requires a signature from upper management before the work can be done. Going all the way to the top might ensure that a way can be found to deenergize the equipment before the work is started.

Job Briefing

In reviewing electrical incidents, there are numerous cases where the worker was not familiar with the equipment, hazards, or procedures resulting in a serious injury or death. A job briefing is essential to familiarize the workers with the task and the boundaries of the task and any specific issues that are unique to the work. It asks the six important questions that need to be answered before work is to begin: Identify, ask, check, know, think, and prepare for an emergency. It requires the workers to identify the hazards, voltage levels involved, and skills required to perform this task. It addresses key questions such as, Does the work need to be performed energized? And is there more that one source of energy? It requires the workers to check drawings, vendor information, and safety procedures. It requires the worker to know what the job is and who is in charge. It develops a line of communication between all affected parties. It prompts workers to think about testing, safety procedures, barriers that may need to be in place, and if they have the right tools for the job. The last step in the briefing form is to prepare for an emergency. Issues such as where is the nearest phone, emergency phone numbers, or fire extinguisher could be the difference between safety and a catastrophe. See Appendix D for a sample Job Briefing and Planning Checklist.

Checklist References

The safe work practices part of both the checklists in Appendix B provides a list of potential tasks that put workers at risk. The checklists provide references to Appendix C for the specific tasks. The tasks have been grouped by the types of equipment or situation that the worker could encounter. Appendix C contains task assessment checklists for many different tasks. These task assessments help walk the worker through a hazard analysis. The checklist requirements are based on specific criteria, such as available fault current and overcurrent protection device tripping time. For tasks outside these parameters, more detailed analyses are required. Chapter 7 describes how to perform a hazard/risk analysis.

Evaluating the Site Assessment

Once the site assessment has been completed, the next step is to categorize the findings. The installation issues need to be corrected to provide a safe installation. The items in the safe work practices section have to be divided into two categories. The first category is potential hazards, such as working on or near energized parts. These items have to be addressed in accordance with Chapters 6, 7, and 8. This will include an analysis that will provide information on the specific hazards, prescribe the hazard boundaries, and assist in the selection of proper personal protective equipment. The other category of the safe work practices addresses procedures and training. These two issues could result in a potential hazard if not addressed. An example would be not having a lockout/tagout procedure or the workers not being trained in the hazards that they may encounter in the normal course of their job tasks. Chapters 4 and 9 provide information on procedures, and Chapter 10 provides information on training.

■ REFERENCES

National Electrical Code® (ANSI/NFPA 70). Quincy, MA: National Fire Protection Association, 2002.

NFPA 70E, *Standard for Electrical Safety Requirements for Employee Workplaces.* Quincy, MA: National Fire Protection Association, 2000.

Chapter 6

Hazard Boundaries

The modern world can be a dangerous place in which to live. Hazards to personal safety are everywhere. People learn to protect themselves from hazards as they grow up, and they learn about boundaries. Children know that a dog might bite, but they soon learn that if a dog is on a strong chain, they need only stay outside the reach of the dog, as limited by the chain, to avoid being bitten.

An electrical hazard can exist with little possibility of personal injury. Transformers and high-voltage electrical installations are surrounded by fences. As long as people stay outside the fence and don't reach inside the fence with an object, they won't be injured. A person must be within the potential reach of the hazard for an injury to occur, Energized, bare conductors exist on towers and poles that distribute electricity to various areas. If a person contacts one of these conductors, he or she could receive a shock or be electrocuted. However, until a person or an object in the person's hand approaches the energized conductor more closely than the arc-over distance, he or she is in no danger from the electrical hazard.

EXPOSURE TO ELECTRICAL HAZARDS

For a shock injury or electrocution to occur, a person must make direct contact with an energized conductor, and current must flow through the body. The direct contact need not necessarily be made with a hand or foot. If a person holds a conductive object in his or her hand and the object contacts an energized conductor, that person essentially is in contact with the energized conductor.

An arcing fault can cause an injury even when the victim is not working on or near an exposed energized conductor. An arcing fault can occur behind a closed door and expel very hot gases and plasma through a door vent, or pressure generated by the fault can force the door to open. However, for an arc-flash injury to occur, a person must be within the ability of the arcing fault to generate thermal energy. Copper vapor or other flying objects cannot cause an injury unless a person is within the travel distance of the vapor (within the boundary).

The hazards of electrical shock and arc-flash burn are very different. Although the potential for injury from shock and arc-flash burn is from the same energy source, people are exposed in different ways. The kind of injury will differ with the hazard. The kind of protective equipment also will differ and must be based on the hazard. Each hazard associated with the work task must be evaluated independently.

The first step in avoiding electrical injury is to know when the potential for injury exists. One way to assess the injury potential is to identify the distance from a potential arcing fault or energized conductor to a person—a boundary that defines the need for protection. Since each hazard must be evaluated independently, the distance within which exposure to injury exists will be different. A boundary must be identified that provides adequate distance from a shock hazard. Another boundary must be identified that will prevent injury from potential arc flash. Approach to an electrical hazard might be from any direction. Therefore, a boundary must include a circular, radial perimeter.

■ BOUNDARIES ASSOCIATED WITH ELECTRICAL SHOCK

People who are trained to be qualified persons should understand the nature of electrical hazards. Qualified persons should know where energized conductors might be and to take extra care when they are near them. A qualified person should know how to avoid contact with the potentially energized conductor. On the other hand, an unqualified person might not be familiar with how or when he or she could be exposed to electrical shock. A boundary must be identified that informs unqualified persons of exposure to electrical shock.

Although qualified persons might have the knowledge to recognize that they are or may be exposed to electrical shock, the degree of exposure is much greater as they approach any energized conductor. As the degree of exposure reaches the point that a person might accidentally contact the energized conductor, another boundary must be identified that will trigger some additional protective measures, such as protective equipment.

Electricity will flow through air when the voltage is sufficiently high and a conductive object is sufficiently close. The point at which current begins to flow through air from a conductor to another conductor is called the arc-over distance. The arc (flow of electricity) flows (over) from one conductive point to another. When a conductive object (such as a hand) is close enough to an energized conductor, the circuit voltage overcomes the impedance of the air, and current flows. When the circuit voltage is high, the arc-over distance can be long. When the circuit voltage is low, the arc-over distance is short. However, an arc-over distance always exists.

Limited Approach Boundary

The limited approach boundary is the approach limit set up for an unqualified person (see Figure 6-1, Illustration of Approach Boundaries). The idea is that a person who does not have the knowledge to recognize when he or she is exposed to electrical shock is not qualified to work any more closely to energized equipment than this dimension allows. The person could unknowingly contact an energized conductor. The point is that the degree of exposure is elevated at any time a person is closer to an exposed energized conductor than the limited approach boundary.

Since maintaining a safe distance between a person and the exposed energized conductor is key, it is necessary to consider that either the person or the conductor could

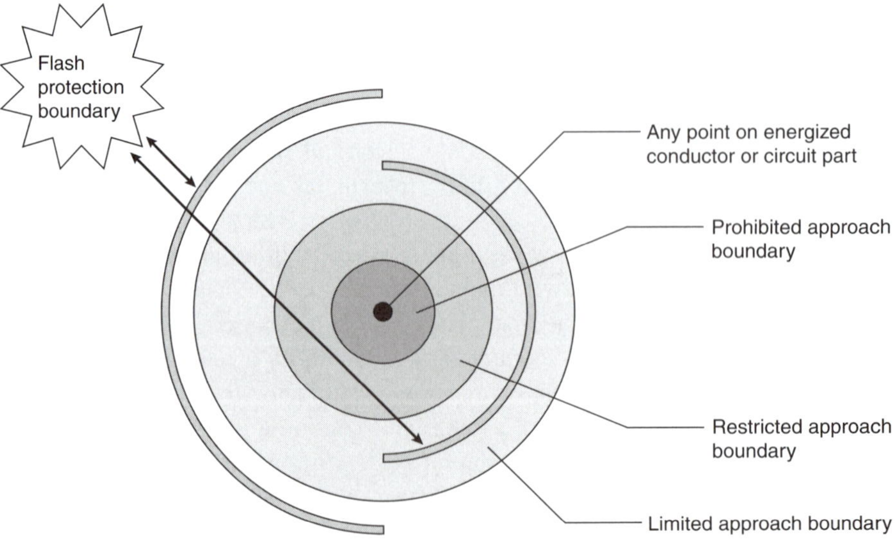

FIGURE 6-1 *Illustration of Approach Boundaries*

move and decrease that distance. The energized conductor could be blown by the wind, or the platform on which the person is standing could move and cause the person to contact an energized conductor. The degree of exposure is greater when the distance between the person and the energized conductor is not fixed. For a person to determine an adequate dimension for the limited approach boundary, he or she must determine if the distance between the worker and the electrical shock hazard is fixed or variable.

If the energized conductor is fixed in place and cannot move, the variable distance from the worker to the energized conductor is controlled by the worker's physical ability to keep clear of the hazard. The limited approach boundary, then, must address the relative difference in the degree of exposure. Table 6-1 identifies a recommended limited approach boundary for various voltage levels.

Table 6-1 also suggests a dimension for the limited approach boundary when the exposed conductor is fixed or movable. As discussed earlier, if a conductor is fixed and the platform on which a person is standing is movable, then the limited approach boundary should be considered to be the same as if the conductor were movable.

The intent of the limited approach boundary is to identify the dimension where increased attention is warranted. Because the boundary is an imaginary surface in air, the issue deals with executing work within space closer than the limited approach. Since the

TABLE 6-1 Approach Boundaries[a] to Live Parts for Shock Protection[b]

Nominal System Voltage Range, Phase to Phase	Limited Approach Boundary[c]		Restricted Approach Boundary[c] (includes inadvertent movement adder)	Prohibited Approach Boundary[c]
	Exposed Movable Conductor	Exposed Fixed Circuit Part		
0 to 50	Not specified	Not specified	Not specified	Not specified
51 to 300	10 ft, 0 in	3 ft, 6 in	Avoid contact	Avoid contact
301 to 750	10 ft, 0 in	3 ft, 6 in	1 ft, 0 in	0 ft, 1 in
751 to 2 kV	10 ft, 0 in	4 ft, 0 in	2 ft, 0 in	0 ft, 3 in
751 to 15 kV	10 ft, 0 in	5 ft, 0 in	2 ft, 2 in	0 ft, 7 in
15.1 kV to 36 kV	10 ft, 0 in	6 ft, 0 in	2 ft, 7 in	0 ft, 10 in
36.1 kV to 46 kV	10 ft, 0 in	8 ft, 0 in	2 ft, 9 in	1 ft, 5 in
46.1 kV to 72.5 kV	10 ft, 0 in	8 ft, 0 in	3 ft, 3 in	2 ft, 1 in
72.6 kV to 121 kV	10 ft, 8 in	8 ft, 0 in	3 ft, 2 in	2 ft, 8 in
138 kV to 145 kV	11 ft, 0 in	10 ft, 0 in	3 ft, 7 in	3 ft, 1 in
161 kV to 169 kV	11 ft, 8 in	11 ft, 8 in	4 ft, 0 in	3 ft, 6 in
230 kV to 242 kV	13 ft, 0 in	13 ft, 0 in	5 ft, 3 in	4 ft, 9 in
345 kV to 362 kV	15 ft, 4 in	15 ft, 4 in	8 ft, 6 in	8 ft, 0 in
500 kV to 550 kV	19 ft, 0 in	19 ft, 0 in	11 ft, 3 in	10 ft, 9 in
765 kV to 800 kV	23 ft, 9 in	23 ft, 9 in	14 ft, 11 in	14 ft, 5 in

[a]All dimensions are the distance from a live part to a worker.

[b]This table is modified from Table 2-1.3.4 in NFPA 70E, *Standard for Electrical Safety Requirements for Employee Workplaces.*

[c]See the Glossary for definition.

degree of exposure to electrical shock is elevated, the person performing a task within the limited approach boundary must recognize and avoid contact with the energized conductor. A qualified person is expected to be able to make that assessment. However, an unqualified person is not expected to have that knowledge and ability. If an unqualified person must perform a task within the limited approach boundary, he or she must be instructed and supervised by a qualified person who has the skill necessary to recognize and avoid contact with any energized conductor.

Since an unqualified person does not necessarily have the understanding to recognize and avoid electrical hazards, then some action must be taken to help the unqualified person know if he or she is approaching an exposed electrical hazard. The training program can instruct the person on how to avoid approaching outside overhead lines closer than 10 feet. However, within a production facility, the distinction between insulated and uninsulated is much less obvious. A physical barricade or adequate warning signs should warn workers that exposed conductors are energized. Unqualified persons should be trained to respect the integrity of the barricades and signs.

Restricted Approach Boundary

As the approach distance to the energized conductor decreases, the degree of shock exposure increases to the point that even qualified persons may not be able to avoid the hazard while performing a task. As the degree of task complexity increases, the degree of exposure also increases. When a worker must perform a task within the space bounded by the restricted approach boundary, he or she should be a qualified person. A qualified person is expected to understand the shock hazard and know how to avoid contact with the exposed energized conductor (see Figure 6-1, Illustration of Approach Boundaries). Since the degree of exposure is elevated by the proximity to the exposed energized conductor, administrative controls should be in place that limit the necessity to perform work in an environment with elevated exposure to shock. The idea is to warn supervisors that a worker is performing a work task that has an elevated risk of shock and electrocution and to know how to mitigate that risk.

The electrical safety program should identify what administrative controls are deemed necessary to accomplish the objective. Consensus standards generally require the following controls:

1. Have a work plan that is documented and authorized by the line organization.
2. Use personal protective equipment (PPE) appropriate for the circuit conditions.
3. Be continually aware of the location of the exposed energized conductor(s).
4. Know how to position his or her body to minimize the chance of incidental contact with the exposed energized conductor.

1. *Have a work plan that is documented and authorized by the line organization.* The work plan should be sufficiently detailed to identify each discrete task that is necessary to accomplish the intended work. The plan could be generated for the specific job, or it could be a standing work plan in the form of an authorized work procedure that is reviewed on a routine basis. The key issue is that the person who executes the work has a written plan that guides his or her actions as the work is executed.

2. *Use PPE appropriate for the circuit conditions.* Before beginning the work, the workers must perform a hazard/risk analysis (see Chapter 7, Hazard/Risk Analysis). The hazard/risk analysis should identify the circuit conditions, including the voltage level to which the worker may be exposed. When work is performed within the restricted approach boundary, workers should wear protective equipment that provides adequate insulating qualities for the voltage that is present. The insulating equipment should have a rating that is at least equal to the voltage that is present (see Chapter 8, Personal Protective Equipment). Generally, voltage-rated protective equipment is referred to as "rubber goods." The term "rated" means that the protective characteristics of the equipment have been established and verified by a third party.

3. *Be continually aware of the location of the exposed energized conductor(s).* As a worker performs a work task, he or she probably focuses attention on the steps that are necessary to physically accomplish the task. As the worker concentrates on the physical activities, the existence and location of the exposed energized conductor might slip from conscious attention. The qualified person must continually re-evaluate the exposure. It might be necessary to assign a second person to observe the actions of the worker and advise him or her when the chance of direct contact with the exposed energized conductor increases.

4. *Know how to position his or her body to minimize the chance of incidental contact with the exposed energized conductor.* When considering a work task, a worker normally looks at the necessary actions and selects a position for his or her body that enables the task to be accomplished with the least amount of physical effort. If an exposed energized conductor is nearby, the worker should first consider how to position his or her body to minimize exposure to the shock hazard. A tool often slips when force is applied. When a tool slips, the worker who is providing the force cannot stop the tool (and his or her hand) from moving in an unintended direction. When selecting a body position, the worker should consider what his or her hand would contact if the tool should slip.

Prohibited Approach Boundary

As a person approaches an exposed energized electrical conductor and the distance gets sufficiently close, the voltage gradient will overcome the insulating value of air, and arc-over will occur (see Figure 6-1, Illustration of Approach Boundaries). Current will flow from the energized conductor to the person. As the voltage increases, the arc-over distance also increases (see Table 6-1). When the voltage is in the normal utilization voltage range, the arc-over distance is so short that direct contact is very likely. Since arc-over probably will occur if the prohibited approach boundary is breached, crossing the boundary has the exact same result as directly contacting the energized conductor. If this type of work is categorized as prohibited work, people are discouraged from performing tasks of this nature.

To avoid shock/electrocution, the person must avoid any potential difference across his or her body. The electron flow theory suggests that current flow can be avoided by insulating the person from the energized conductor or by insulating the person from contacting any point energized at a voltage different from the exposed energized conductor. The most common point that is at a different potential is earth ground.

Rubber protective gloves, rubber insulating sleeves, and rubber insulating blankets that serve to insulate a worker from an energized conductor are readily available. Insulating the person from the energized conductor is a reasonable choice. However, using protective footwear and matting intended to insulate the worker from earth ground is a very poor choice. Although it is realistic to isolate the worker from ground in a transmission line setting, the choice is not practical in an industrial setting and should not be used.

The electrical safety program should take additional steps to decrease the potential for injury from the energized conductor. The primary reason for the additional steps is to ensure that no real alternative way to perform the work is possible. Most of the time, in an industrial setting, an electrically safe work condition can be established by scheduling the task at a time when the process demands are reduced. Only in rare instances is it necessary to perform work within the prohibited approach boundary. When work must be performed within the prohibited approach boundary, the following steps must be taken:

1. Train the worker to work on specific energized conductors.
2. Document a specific plan that justifies the need for energized work.
3. Perform a hazard/risk analysis.
4. Obtain management authorization accepting the justification, the work plan, and the hazard/risk analysis.
5. Use PPE that is rated for the voltage and energy levels that are available at the work location.

1. ***Train the worker to work on specific energized conductors.*** When prohibited work must be done, the worker must be trained specifically to perform work directly on energized conductors. The intent of the specific training is to avoid any chance of an unexpected arrangement of wiring and devices while the work task is being performed. The training should include a demonstration that the worker understands how he or she is exposed to an electrical hazard. The worker must know how to avoid initiating a fault and have the physical ability and skill to act as necessary to accomplish that objective.

2. ***Document a specific plan that justifies the need for energized work.*** Prohibited work tasks involve risks that are great. Each step that is necessary to execute the work task should be written in a plan. It is important that no step be left out. After the written plan is completed, it should be reviewed and critiqued by at least one other qualified person and by the worker's immediate supervisor.

The plan should include a section that shows that the worker requested a shutdown of the equipment while the work is performed (see Appendix E, Energized Electrical Work Permit). This section should provide a detailed reason why the equipment or circuit cannot be de-energized while the work is performed. This section also should contain the signature of the person responsible for the equipment or circuit who accepts the increased risk associated with working on or near an exposed energized conductor.

The worker must agree that the work task can be performed safely as detailed in the written plan. It is possible to perform work directly on an energized conductor. However, the degree of risk is significantly elevated. The elevated risk warrants increased controls, planning, and authorization. Everyone must agree that the prohibited work is necessary, that the plan is complete, that it includes all relevant information, and that all parties accept the elevated degree of risk.

3. ***Perform a hazard/risk analysis.*** Work performed on or near exposed energized conductors is dangerous. Before a work plan can be generated, performing a hazard/risk analysis is necessary. See Chapter 7 for more information. If the voltage level is more than 50 volts, the risk of shock or electrocution is great. As the voltage level increases, the degree of risk increases. Each step of the work plan should be analyzed to ensure that the exposure is mitigated to the greatest extent possible. Depending on the amount of energy that the electrical source can supply, a significant arc-flash hazard might exist. In all instances with potential for an arc-flash event, the potential for an arc-blast also exists. An arc-flash thermal hazard cannot exist without a companion arc-blast event. As the circuit energy level increases, the degree of the arc-flash hazard also increases.

After the hazard/risk analysis has been completed, the worker must review each step of the plan with an eye toward reducing either exposure or risk. He or she might need to revise the plan based on the result of the hazard/risk analysis.

4. *Obtain management authorization accepting the justification, the work plan, and the hazard/risk analysis.* Regardless of whether or not the plant or facility manager or the owner accepts responsibility, he or she is directly responsible for any injury that might occur on the site. In many instances, the top-level manager is held accountable for all injuries that happen on site. It is important, then, that the plant or facility manager is the person who authorizes the prohibited work. Being asked to "sign off" on a work plan that exposes a worker to an elevated risk of electrocution causes significant soul-searching by the manager. Frequently, a way can be found to de-energize the equipment or circuit to do the work before reaching the step that seeks permission from the manager.

5. *Use PPE that is rated for the voltage and energy levels that are available at the work location.* The hazard/risk analysis indicates what hazards are associated with the work task as well as the degree of the hazard. Once that information is known, the qualified worker then knows what protection is required. This equipment must match the protective characteristics of the PPE with the degree of hazard to which the worker will be exposed (see Chapter 8 and Appendices H, I, and J for more information). After the PPE is identified, the worker should again review the work plan to determine if each step can be accomplished while wearing the necessary PPE. If not, then the work should not be performed until a different plan is generated.

ARC FLASH

Only one boundary needs to be considered when the hazard is arc flash. If a chance of injury exists, then it is best to take some step to mitigate that chance for injury. Consensus standards consider that a chance for injury exists if sufficient energy reaches the skin surface to result in a second-degree burn. The term "curable burn limit" is sometimes used to describe the onset of a second-degree burn. Electrical safety literature suggests that if 1.2 calories per square centimeter of thermal energy is received by a person's skin, he or she probably will receive a second-degree burn. In theory, one can evaluate the amount of energy available at the point in the circuit where an arcing fault occurs that will release enough thermal energy to reach the generally accepted burn limit at a predefined distance. If a person is closer than this predefined distance, he or she could receive a second-degree burn. As long as the person keeps outside the predefined distance, the chance of receiving a second-degree burn is small.

Flash Protection Boundary

A predefined distance that might result in a second-degree burn is called the flash protection boundary. If an arcing fault releases enough energy to cause a burn at some distance, it stands to reason that the closer a person is to the arcing fault, the greater the amount of available thermal energy will be.

It also stands to reason that if the temperature of the arc plasma increases, the arc will release more thermal energy into the surrounding environment. During a faulted condition, the current in the circuit increases until the energy source cannot provide any greater current. The change in fault current is determined by the electrical parameters of the circuit. For instance, the inductive component of the circuit tends to limit the rate at which the fault current changes. The overcurrent protective device limits the duration of the fault current. The inductive component of the circuit is not easily modified. However, an overcurrent protective device (fuse or circuit breaker) can and should be selected based on the ability of the device to limit the duration of any fault current. The critical issue is the ability of the device to protect people.

For a burn to occur, a person's skin must be exposed to an elevated temperature long enough for the injury to occur. If the surface of a person's skin is elevated to 176°F for 0.1 second, a second-degree burn probably will occur. If the energy source is an arcing fault, the duration probably will be less than one second. However, if the person's clothing begins to burn, the duration of the exposure will be much longer. The exposure probably will be several seconds or perhaps even several minutes. Although the temperature of burning clothing is much lower than the temperature of an electrical arc, the exposure duration will probably result in a more significant burn injury. Some clothing is constructed from materials that will melt and then ignite. Burn centers relate that a burn injury with melted fabric imbedded in the burned skin tissue is particularly difficult to cure. If a person is closer to a potential arcing fault, he or she must take some steps to mitigate the potential arc-flash injury.

When compared with the temperature of an electrical arc, the ignition temperature of clothing is relatively low. Depending upon the material of construction, clothing may ignite when its temperature reaches between 700°F and 1400°F. The temperature of burning clothing will be on the order of 1400°F. An injury from exposure to burning clothing is probably going to be very severe.

Mitigating Exposure to Arc Flash

Since the temperature within an arcing fault can approach 35,000°F, any person who might be exposed to this hazard must select and wear protective equipment that will mitigate the exposure. The protective equipment should have characteristics that enable it to resist ignition, and the equipment should not melt under normal conditions. The protective equipment might be clothing, a face shield, a hood with a viewing window, gloves, or shoes.

The degree of the shock hazard is dependent on well-defined physical characteristics of electron movement. Ohm's law determines the flow of current, but the degree of the thermal hazard is not easily predictable. The laws of physics still apply; however, the physical laws governing both the flow of electricity and the transfer of heat also apply. The thermal energy from the arcing fault to the surrounding environment is transferred by conduction, convection, and radiation. The arcing fault radiates energy, and some of that energy might be received by a nearby person. As the arc heats the air in the immediate vicinity of the arc, the pressure wave that is simultaneously created transfers some of the thermal energy to nearby surfaces by convection. As a mechanism of transferring thermal energy during an arcing fault, conduction only comes into play when the receiving material is the metal conductor or other metal components that are nearby.

To interfere with the transfer of thermal energy from the electrical arc to a person who might be nearby, a thermal insulator must be inserted between the electrical arc and the person's skin or clothing. The thermal insulation normally takes the form of clothing or other thermally protective material, such as a face shield and gloves.

The American Society of Testing and Materials (ASTM) is the standards organization that provides consensus information on how to determine the thermally protective characteristics of clothing. ASTM 1959, "Standard Test Method for Determining the Arc Thermal Performance Value of Materials for Clothing," is a national consensus standard that defines a test procedure intended to explore the protective nature of protective clothing. ASTM 1959 establishes standard nomenclature that describes the amount of heat energy in calories per square centimeter that can be received on the surface of the clothing without conducting enough thermal energy to cause a second-degree burn. Clothing that passes the test criteria described in ASTM F 1959 is assigned an arc thermal protective value (ATPV) in calories per square centimeter. If the insulating nature of the insulating material is sufficient to prevent a second-degree burn, it will prevent flammable clothing from igniting.

If the amount of thermal energy generated by an arcing fault could be predicted in calories per square centimeter, protective clothing that has a higher protective value could be worn. That protective clothing would prevent a significant burn injury from occurring. Exposure to a potential arc-flash hazard would be controlled to an acceptable level.

As the amount of fault current increases, *logic* suggests that the degree of the hazard would also increase and, as the fault current decreases, the degree of hazard would decrease. Neither of these *logical* statements is necessarily true. The amount of arc current is one significant factor. However, the clearing time of the overcurrent device is the more significant factor. Sometimes, an increase in the fault current will reduce the clearing time of the overcurrent device with a resultant decrease in the degree of exposure. If the overcurrent device is current limiting and the fault current is below the lower limit of current limitation, the clearing time will be much greater and result in a significant increase in the degree of the arc-flash hazard.

Determining the Degree of the Arc-Flash Hazard

No completely reliable method exists that will accurately predict the amount of energy released in an arcing fault. A significant amount of testing has been performed. The testing has relied upon the test setup and test method described in ASTM 1959, which is designed for testing clothing. Most methods attempt to determine the amount of thermal energy that would be generated in an arcing fault by measuring electrical parameters that generated the thermal energy on which clothing is rated.

Several power system design and simulation software programs contain an algorithm that can predict the amount of thermal energy in calories per square centimeter. Some computer programs calculate and predict only the thermal intensity of an arcing fault. The National Fire Protection Association's NFPA 70E, *Standard for Electrical Safety Requirements in Employee Workplaces,* contains a table that identifies several work tasks and then assigns a hazard category for the work task. The standard then helps the user to select protective clothing based on the assigned hazard category.

Most of the calculation methods are based on defined electrical circuit parameters such as the following:

- Bolted fault
- Available short-circuit current
- Fault-clearing time
- Voltage level
- Distance from the arc to the worker
- Physical properties associated with the conductor (whether or not the arcing fault is in an enclosure with a door open, in open air, or contained within a closed enclosure)

The various methods of predicting the degree of the arc-flash hazard do not correlate at all energy levels. However, each of the methods do correlate at one or more points as the amount of available energy changes from a low level to a high level.

Table 3-3.9.1 from NFPA 70E (see Appendix G, Hazard/Risk Category Selections) seems to be the most widely used method of selecting protective clothing. It is important to recognize that the NFPA 70E table does not provide 100-percent assurance that an injury will be avoided. It is also important to recognize that no method of selecting protective clothing will provide 100-percent effective protection from a thermal injury.

The parameters that impact the thermal hazard vary widely from one exposure to another. For instance, an enclosure tends to direct the thermal energy when the arcing fault is in an empty enclosure. However, other components within an enclosure, such as a disconnecting device or a contactor, change the reflecting nature of the enclosure. In addition, the following factors can also change the impact of a hazard:

- The color of the enclosure changes the energy-absorbing characteristics of the enclosure.
- The materials of construction for the electrical conductor impact the amount of energy that is converted to thermal energy.
- The distance between the arcing fault and the worker dramatically changes the amount of energy received by the worker.
- Variation in the distance between the arcing terminals as the conductor vaporizes changes the amount of energy converted into thermal energy.
- The maintenance condition of the overcurrent device changes the amount of available electrical energy.

Given all of these factors, predicting the amount of energy that could be received by the worker's skin or clothing is extremely difficult.

The organization should consider selecting one method of predicting the degree of the thermal hazard associated with an arcing fault. The selected method should be based on the type of exposure that exists on a facility. Equipment on the site should be labeled with sufficient information for a worker to select protective clothing that will mitigate exposure. Since all methods indicate that an injury is still possible, the authors recommend that any error be on the side of more protection.

After the members of the organization select a method of predicting the degree of arc-flash hazard, they must administer the protective equipment. If the protective clothing required by the organization is more restrictive than the workers are willing to accept, administering the program will be difficult. Employees sometimes will more easily accept the selected protective clothing if it is less restrictive rather than more restrictive.

Where Arc-Flash Exposure Exists

Electrical equipment that is installed in accordance with the *National Electrical Code®* (*NEC®*) and maintained in as-installed condition is considered safe from exposure to shock, unless the enclosure is abridged for some reason. However, a person may be exposed to an arc-flash injury even if the installation meets the requirements of consensus standards.

Some electrical equipment requires ventilation to remove heat that is generated during normal operation. If the electrical equipment has ventilation louvers or air filters on its external surface, the pressure build-up during an internal arcing fault will probably force gases and heated air out through the ventilation opening. If a person is standing nearby, the person could be injured by this heated air and vapor. The plasma from the arc might exit through the ventilation openings. If a device has been added to the door since installation, or if a device has been removed and the space temporarily covered, the components that were added to the door may be expelled with significant force.

Listed and labeled electrical equipment is tested to withstand magnetic force associated with a bolted fault. However, only arc-resistant electrical equipment is tested to withstand the forces generated by an arcing fault. The forces generated in an arcing fault frequently can cause doors to open unless the equipment is arc resistant. The authors recommend arc-resistant electrical equipment to avoid unnecessary exposure when moving the handle of switches or circuit breakers.

■ REFERENCES

ASTM 1959, "Standard Test Method for Determining the Arc Thermal Performance Value of Materials for Clothing." Philadelphia: American Society for Testing and Materials, 1999.

National Electrical Code® (ANSI/NFPA 70). Quincy, MA: National Fire Protection Association, 2002.

NFPA 70E, *Standard for Electrical Safety Requirements for Employee Workplaces.* Quincy, MA: National Fire Protection Association, 2000.

Chapter 7

Hazard/Risk Analysis

For many years people have known that electricity can be dangerous. Although the knowledge of electricity has increased greatly through the years, electrocution remains one of the leading causes of fatality in the workplace. Understanding and recognizing the hazards of electricity are the key to preventing injury. Those who work with electrical equipment today must be highly trained specialists who know the hazards involved and how to avoid those hazards.

To avoid unnecessary exposure to electrical safety hazards, workers must identify all of the hazards that they might encounter while performing a work task. Depending on the complexity of the work task, the hazard/risk analysis might be either simple or complex. For simple work tasks, the analysis need not even require a written document. But the analysis can become complex when it involves complex work tasks, complex electrical circuits, complex electrical equipment, or a complex process. The result of the analysis will identify all hazards that are associated with the work task.

After the worker has identified all the hazards, he or she must then determine the degree of risk associated with each hazard. The existence of a hazard does not necessarily expose a worker to an injury. As the degree of risk increases, however, the chance of an injury also increases. For example, for a shock injury to occur from an exposed electrical conductor that is energized at 480 volts, the worker must be closer than the *arc-over distance* for 480 volts. At this voltage level, if the worker approaches the "live part" with a body part or a tool, then he or she probably will be injured. Therefore, the analysis must consider the chance that the worker will interact with the hazard.

After the possible hazards have been identified and the chance of an energy release associated with each hazard has been evaluated, determining what actions or equipment might be taken to eliminate or mitigate exposure to each hazard becomes necessary. In the example above, the worker should first choose to shut down the process and create an electrically safe work condition. Next, the worker might choose to wear insulating gloves that are rated at least for the highest voltage to which he or she will be exposed. In this example, Class 00 gloves would satisfy the need for protection from shock. However, other electrical hazards might also exist. The voltage-rated gloves might provide protection from one hazard but not from another. The worker must consider each hazard independently and then merge the choice of protective equipment with consideration of other hazards to reach a satisfactory conclusion.

◼ ELECTRICAL HAZARDS

Electrical energy is relatively easy to transport from a source to where it will be used. All that is needed is a generator, a complete circuit, and a way to use the energy. Electrical energy normally is selected to provide the power necessary to drive equipment. In its natural state, electrical energy is not usable. Electrical energy is easily converted to another form of energy. For instance, an electrical heater converts electrical energy into thermal energy for warmth. An electrical motor converts electrical energy into mechanical energy that can be used to turn a pump or pull a belt. Because electrical energy is easily converted, it is widely used. The broad use of electricity tends to result in acceptance and also apathy by the user community. The apathy results in not considering the hazards that are associated with the use of electrical energy.

Fire

Almost everyone understands that electricity can cause a fire. Forensic analyses frequently find that fires were started by the electrical system. Many times, the analysis is accurate.

When electrical current flows through a conductor having resistance or impedance, some energy is converted into heat by the resistance or impedance of the conductor. When the heat energy reaches the ignition temperature of adjacent material, the material ignites, and a fire is started.

Small conductors have more resistance than large conductors. Controlling the amount of current in the circuit controls the amount of heat generated by the conductor resistance. Limiting the current in the circuit to the ampacity assigned by NFPA 70, the *National Electrical Code*® (*NEC*®), will avoid igniting a nearby material.

Electrical connections (splices or terminations) heat and cool as the amount of current flow changes; equipment is turned on and off in the normal course of use. The conducting metal heats and cools with each on-off cycle. The heating and cooling cycle sometimes results in slight stretching of the conductor material and, therefore, decreases the contact pressure in the connection. Decreased contact pressure decreases the ability of the junction to carry electrical current safely, and the resistance of the joint increases. As the resistance increases, losses (heating) increase. Flammable material may be ignited by the thermal energy generated in the high-resistance connection.

Manufacturing processes frequently use flammable materials. The flammable materials may be vapor or dust. Material that is not flammable in its normal state might be explosive as a dust or powder. Prevention of an explosion of a gaseous material or dust requires action to be taken that will remove at least one leg of the fire triangle: fuel, oxygen, and heat. Any time an explosive atmosphere is permitted or required to exist, ignition could occur from an electrical energy source. The electrical energy source could come either from the installed electrical system of conductors and equipment or from a discharge of static electricity.

Chapter 5 of the *NEC* defines requirements that, if properly applied, will prevent such an explosion from the installation. However, processes and actions that are neither governed by the *NEC* nor affected by its requirements can generate static electricity. An explosion hazard can exist even if all *NEC* requirements are satisfied.

The *NEC* defines installation requirements that can avoid fire in a new installation. As a facility ages, however, equipment deteriorates. Implementing steps to abridge the equipment deterioration then becomes necessary. NFPA 70B, *Recommended Practice for Electrical Equipment Maintenance,* suggests steps that will help to overcome the deterioration process.

Shock

The human body's muscle system can be considered to be an electrical system. A muscle contracts and moves a body part when the muscle receives an electrical signal. An electrical current is chemically generated in the brain and transmitted to a specific muscle as electrical current through a system of conductors (nerves). When the current reaches the intended muscle, the muscle interprets the current flow as a signal to contract.

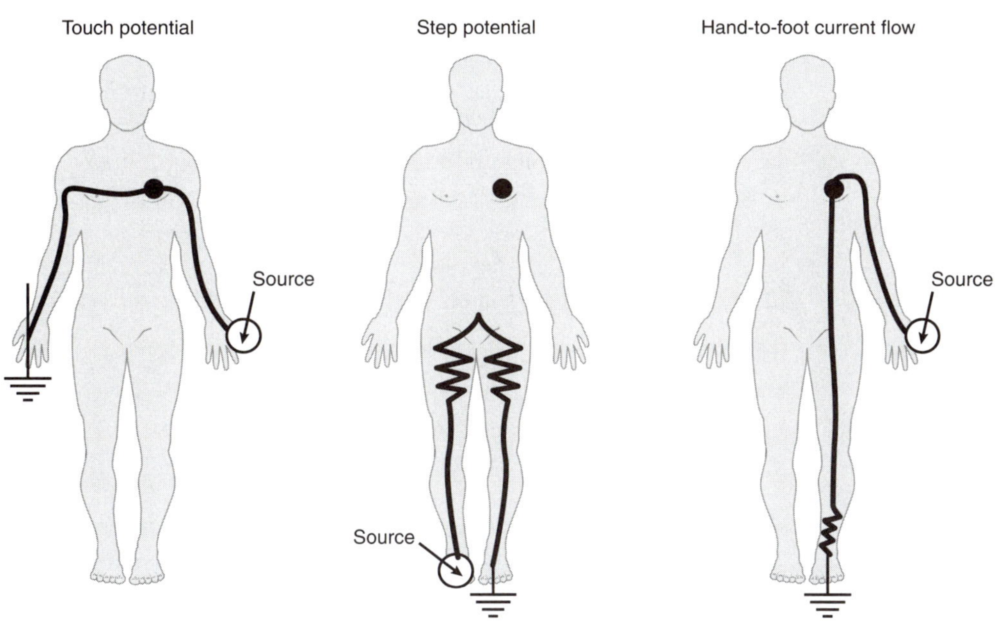

FIGURE 7-1 *Current Flow through the Human Body*

Electrical shock occurs when electrical current flows through a body part. When an externally generated current flow of sufficient magnitude reaches a muscle, the muscle does not recognize that the current was not generated by the brain and acts accordingly. The muscle tissue contracts, and it remains in the contracted state until the external source of current is removed (see Figure 7-1, Current Flow through the Human Body).

Ohm's Law applies even when the conductor material is human tissue (see Figure 7-2, Illustration of Ohm's Law). Blood contains nutrients and chemicals. The chemicals tend to make the blood a very good conductor of electricity. The function of nerve tissue is to

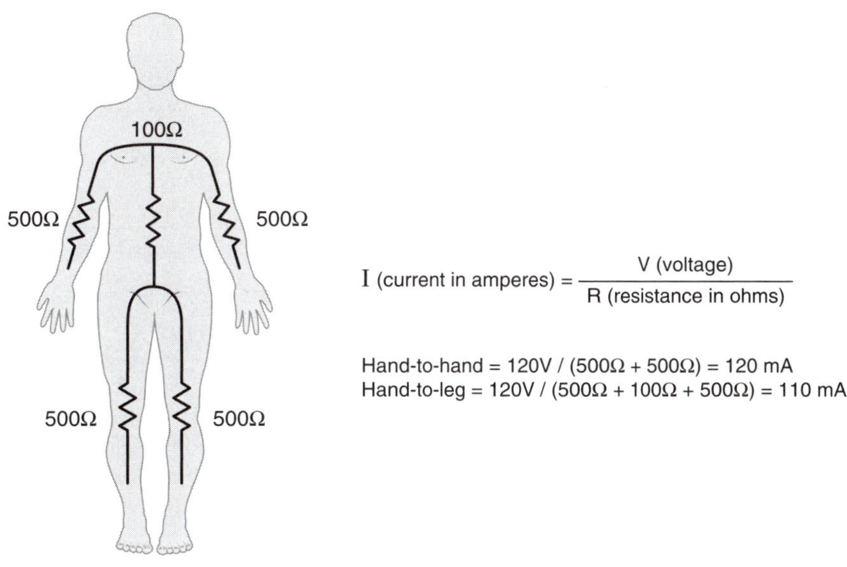

$$I \text{ (current in amperes)} = \frac{V \text{ (voltage)}}{R \text{ (resistance in ohms)}}$$

Hand-to-hand = 120V / (500Ω + 500Ω) = 120 mA
Hand-to-leg = 120V / (500Ω + 100Ω + 500Ω) = 110 mA

FIGURE 7-2 *Illustration of Ohm's Law*

conduct electrical signals to various parts of the body, so nerve tissue is an excellent conductor. Fortunately, neither blood nor nerve tissue normally is exposed directly to contact with an external source of current. For an external current to reach these tissues with low resistance, the current must flow through the skin, overcoming the skin's resistance.

Although internal resistance varies with many factors, including the current density in the conducting tissue, most sources accept 100 ohms as the internal resistance of a human body.

Several factors can impact the contact resistance of the skin's surface with any external current source, such as these examples:

- Coarseness of the skin
- Area of the contact
- Contact pressure
- Degree of wetness of the surface

The resistance of the surface contact will vary from a couple of hundred ohms to several hundred ohms as these factors change. Most authorities accept 500 ohms as a reasonable expectation for the contact resistance, unless the skin's surface is broken, thus exposing the external surface to blood. When the skin's surface is broken and internal fluids are exposed to the contact surface, any contact resistance and any skin resistance effectively disappears. Only internal body resistance remains.

As suggested by Figure 7-2, the resistance of the circuit is 500 ohms, plus 100 ohms, plus another 500 ohms. If a person is working on an electrical circuit that is energized at 120 volts ac and one of the person's hands makes incidental contact with the energized conductor, 120 volts is impressed from one hand to the other. Applying Ohm's Law to this typical circuit [120/(500+100+500)], shows that 110 milliamps of current will flow. If a person's hands are wet or in perspiration-soaked gloves, the contact resistance will decrease, and the amount of current will increase. If the person's hand is dry or wearing a clean, dry glove, the current will decrease.

Table 7-1 illustrates how a person's body might react to electrical current flow. The body's reaction is time-dependent on the duration of the current flow. Not only does the current increase with time, the body reaction is more severe.

TABLE 7-1 Reaction of Human Body to Electric Current[a]

Effect of Current	[b]AC Current in Amperes —Large Bodies	[b]AC Current in Amperes —Small Bodies
Perception threshold (tingling sensation)	0.0010	0.0007
Slight shock—not painful (no loss of muscle control)	0.0018	0.0012
Shock—painful (no loss of muscle control)	0.0090	0.0060
Shock—severe (muscle control loss, breathing difficulty)—onset of "let-go" threshold	0.0230	0.0150
Possible ventricular fibrillation (3-second shock)	0.1000	0.1000
Possible ventricular fibrillation (1-second shock)	0.2000	0.2000
Heart muscle activity ceases	0.5000	0.5000
Tissue and organs burn	1.5000	1.5000

[a]From *Electrical Safety in the Workplace,* by Jones and Jones.
[b]U.S. Department of Energy Electrical Safety Guidelines, Appendix A, 9/93.

As indicated in Table 7-1, when the amount of current approaches 20 mA, the let-go threshold is reached. When a muscle receives an electrical signal, it interprets the signal to mean it must constrict. When the signal reaches the degree at which internal processes cannot overcome the external source of energy, the person cannot override the signal to constrict. The let-go threshold has been reached. The person can escape only with external intervention in the incidental circuit.

If the current flow increases to approach 100 mA, involuntary muscular action is impacted. Although still based on the duration of the exposure, ventricular fibrillation is likely within about 1 to 3 seconds. As suggested by Table 7-1, the amount of current necessary to cause catastrophic consequences is very small. The table illustrates that the difference between a small shock that causes little consequence and ventricular fibrillation is very small.

Arc Flash

Historically, electrical equipment is tested and rated on the basis of its ability to withstand the magnetic force that results from current flow. The magnetic force is a normal (and sometimes useful) result of electrical current flowing in a conductor. The magnetic force generates a mechanical force that the conductor must resist by mechanical bracing or support.

An electrical arc results when a conductor is impressed with a voltage while it is in contact with another object that cannot handle the full short-circuit capacity of the energy source. The object that initiated the short circuit is either vaporized by the current flow or moved by the mechanical force. After the short circuit occurs, the tendency is for the current to continue to flow through the conductive plasma until some circuit element removes the voltage or the circuit condition breaks the current flow.

Any time an electrical circuit is broken, the inductive characteristic of the circuit will not allow an immediate current change to zero. The current attempts to keep flowing, resulting in an electrical arc. Most electrical switches and contactors contain arc chutes that are intended to help quench the arc, as well as to provide isolation between poles. As long as the arc is kept from connecting with another arc, pole, or ground, the arc will dissipate without difficulty. However, if the switch or contactor is exposed to current that is beyond its capacity to break, an arcing fault will occur.

Normal heating and cooling may loosen joints in electrical conductors. As the joint becomes loose, the inherent resistance in the joint increases. The increased resistance results in increased heating. When the thermal temperature of the joint reaches the melting point of the conductor material (normally copper), the resulting vapor will initiate an arcing fault. The arcing fault probably will migrate to a multiphase fault and continue until the overcurrent device clears the fault.

When the conductor is metallic vapor or arc plasma, the electromagnetic force tends to act on the conducting medium and cause the electrical arc to move. When the arc moves, the parameters (such as path of current flow) associated with the arc change. When the voltage is from an alternating-current source and the voltage cycles below the voltage necessary to maintain the current flow, the arc extinguishes. The arc could be reestablished if the voltage is able to overcome the impedance again.

When plasma exists due to the flow of electrical current, the energy dissipated in the arc plasma is converted from electrical energy to other forms of energy, such as the following:

- Thermal energy
- Acoustic energy
- Kinetic (pressure) energy
- Electromagnetic energy

Electromagnetic energy is not restricted to any single bandwidth of the spectrum. Some energy is in the visual band, some is in the infrared band, some is in the ultraviolet band, and so on. Each energy type has characteristics that are or may be hazardous to human tissue.

In most standards and technical papers, arc flash is considered to be almost synonymous with the thermal hazard associated with the heat generated by the arc. With sufficient energy and time available, the arc temperature will reach 35,000°F unless the source of electrical energy is removed. Although the duration of the arc limits the temperature, the maximum arc temperature may be reached in a matter of cycles rather than seconds. Conductors melt and vaporize at much lower temperatures.

The heat energy generated by the arc is consumed by heating components in the immediate vicinity (if the arc is enclosed) or heating any other nearby device or person. The thermal energy is transmitted from the arc through the normal thermal processes of conduction, convection, or radiation. Arcing faults in electrical equipment frequently reach 14,000 to 16,000°F. Any person who happens to be positioned in the vicinity of such an arc will receive some of the thermal energy that is transmitted from the arc.

The amount of energy received by the exposed person is governed by several factors. The arc temperature begins at the normal temperature of the components and rapidly increases. In fact, the rate of temperature rise seems to increase with time. If the overcurrent device breaks the current flow, the temperature increase will stop. The amount of energy received by the exposed person, then, is dependent on the duration of the arc.

The amount of energy received on any surface exposed to an arc depends on the distance from the arc to the receiving surface. Surfaces that are near to an arc receive more energy than those farther away. One protective strategy, then, is to maintain space between the arcing fault and the worker. Unfortunately, a worker must normally have his or her body within arm's reach to perform a task. Since the worker must be nearby, using some nonflammable protective equipment may be necessary to avoid injury.

A second-degree burn occurs in human skin tissue when the temperature of the skin is raised to 175°F for 0.1 second. To avoid a second-degree burn, the skin must be protected from reaching that temperature. Any surface that is exposed to a temperature of many thousands of degrees will be heated quite rapidly.

Should a worker's clothing ignite or melt onto the skin, the degree of potential injury is very great. Nylon, polyester, and similar synthetic materials tend to melt when exposed to extreme heat. The time necessary to remove burning clothing exposes the person's skin to a very high temperature for a relatively long period of time. The best way to eliminate exposure to a burn injury is to create an electrically safe work condition (see Chapter 9, Lockout/Tagout).

If eliminating exposure to a potential arcing fault is not possible, the worker needs to protect himself or herself from the associated thermal hazard that might result from an arcing fault. Chapter 6, Hazardous Boundaries, discusses how to determine the degree of the potential thermal hazard.

Arc Blast

Most electrical conductors are copper, especially in equipment. During the arcing fault, a copper conductor will first change from solid to liquid, then from liquid to a gaseous state. This process of changing state modifies the amount of physical space required for the copper molecules.

When water is changed from liquid to gas (steam), the water molecules require about four times as much space if the pressure is not held constant. Similarly, when copper changes from liquid to gas, the copper molecules expand several thousands of times. The surrounding air (in open air) offers some resistance to the expanding gas, causing the pressure to increase. Sometimes, equipment is destroyed and parts are expelled in any open direction. The copper vapor also is expelled in any open direction. This situation results in two distinctly different hazards. A person can be injured when a part expelled by the pressure increase makes contact with the person. Copper vapor that is expelled from the arc can be inhaled and result in a thermal injury to the lungs.

If the arc is contained within an enclosure, the pressure will likely increase until the enclosure fails. Sometimes, vents and filters intended for internal cooling are installed on the door. The vents and filters provide an opening to the outside. Should an arcing fault occur within the enclosure, the products of the combustion probably would be discharged through the vents or another opening. Any unprotected person who is near the vent discharge might be injured.

Electrical equipment normally is tested and rated to withstand the magnetic forces generated by a bolted fault. The integrity of the enclosure can be expected to remain intact in the event of a bolted fault. During a bolted fault, the fault energy is dissipated through the circuit. During an arcing fault, almost all of the fault energy is dissipated in the arc. The pressure that results from the arc probably will damage the integrity of the enclosure.

Arc-resistant electrical equipment avoids any opening in the front of the equipment and generally is designed to direct the blast and combustion products away from the front of the equipment. Any person standing in front of arc-resistant equipment would not be injured, provided the equipment door(s) is (are) closed as intended by the manufacturer. When a maintenance or repair task is being performed, arc-resistant equipment is no longer arc resistant. A fault initiated by a worker will direct the arc energy at the worker just as equipment that is not arc resistant. However, the authors strongly recommend arc-resistant electrical equipment for all future installations.

Other Safety Hazards

While performing a task associated with electricity, a person may be exposed to several other types of hazards. Injuries from falls and falling objects are one of the most common types of injury. Although falling objects frequently result from unstable storage, a person could have dropped the object. When planning how to perform a work task, the worker should take steps to minimize his or her exposure to injury from falling from a platform or dropping objects onto a colleague below. If a work area is organized and free from clutter, exposure to tripping also is reduced.

When a worker performs a hazard analysis, he or she must consider all hazards associated with executing the task and take actions to minimize his or her exposure. If exposure to the hazard is not eliminated, then the worker must take some action to reduce the risk of injury. In most instances, personal protective equipment (PPE) can reduce the risk of injury. However, the worker must select the PPE after understanding all the hazards and how the PPE might impact exposure to another hazard.

For instance, a face shield that provides excellent protection from impact by a foreign body might increase the risk of injury from a thermal hazard. Likewise, a face shield that absorbs ultraviolet energy of an electrical arc might increase the chance of falling from a platform.

PERFORMING THE HAZARD/RISK ANALYSIS

For an injury to occur, all of the following conditions are required.

- A hazard must exist.
- A person must be exposed to the hazard.

- A release of the energy contained within the hazard must occur.
- The degree of hazard must exceed the protection offered by the PPE used by the worker.

The objective of a hazard analysis is to determine if any of the above conditions exist and, if one or more of the conditions exist, what action might be taken to eliminate or reduce the degree of injury risk. Some risk might be acceptable. Although driving or riding in an automobile exposes the person to risk of a wreck, for instance, most people consider the degree of risk to be reduced to an acceptable level by maintaining the automobile in good condition, wearing a seat belt, and being alert to other vehicles on the road. Although opening a door or removing a cover to determine the conditions within electrical equipment might expose a person to some risk of injury, adequate PPE might decrease the risk of injury to an acceptable level. The result of the hazard/risk analysis provides information only about the work task. A person must then determine if any risk identified in the process is acceptable.

To execute a hazard/risk analysis, a person must think about each discrete step necessary to execute the intended job and then consider the safety hazards associated with each step. The person must think first about all of the work tasks and then think about the safety hazards associated with each work task.

The flow diagram in Figure 7-3 illustrates a thought process that can be applied to each step in the work process. Although readers may think that the diagram appears complex, after they apply the questions in the flow diagram a couple of times, they find that determining a result for each step in the work plan becomes relatively simple. The result of the thought process contained in the diagram will be an understanding of the potential for injury.

Using the Flow Diagram

Block 1. In blocks 1, 1a, and 1b, the objective is to determine if an electrical hazard exists. General consensus suggests that 50 volts is the potential level at which a shock hazard begins. Although most consensus standards, as well as OSHA 29 CFR 1910, Subpart S, suggest that voltages less than 50 volts are not electrically hazardous, it is important to understand that only shock and arc-flash hazards are considered. Other hazards do exist and should be considered. In many instances, PLC- and DCS-conductors are energized below 50 volts. It is sometimes possible to interrupt a process that could result in an incident of major proportion. The flow diagram does not consider that batteries can store large quantities of energy with large arc-flash boundaries. It is very important to understand what is meant by the answer to each question in the diagram. If a hazard exists, proceed to block 2.

Block 2. In blocks 2, 2a, and 2b, the objective is to determine the degree of shock hazard. As the voltage level increases, the *degree* of the shock hazard also increases. In other words, contact with 480 volts is more likely to result in an electrocution than contact with 115 volts. By understanding the degree of the hazard, it becomes possible to select an appropriate procedure, authorization, and/or protective equipment. Proceed to block 3.

Block 3. In blocks 3, 3a, and 3b, the objective is to determine the degree of arc-flash hazard. At this point in the thought process, the arc-flash boundary dimension is required (see Chapter 6, Hazard Boundaries). Of course, as the quantity of available energy increases, the potential for injury from arc flashes also increases. With this information,

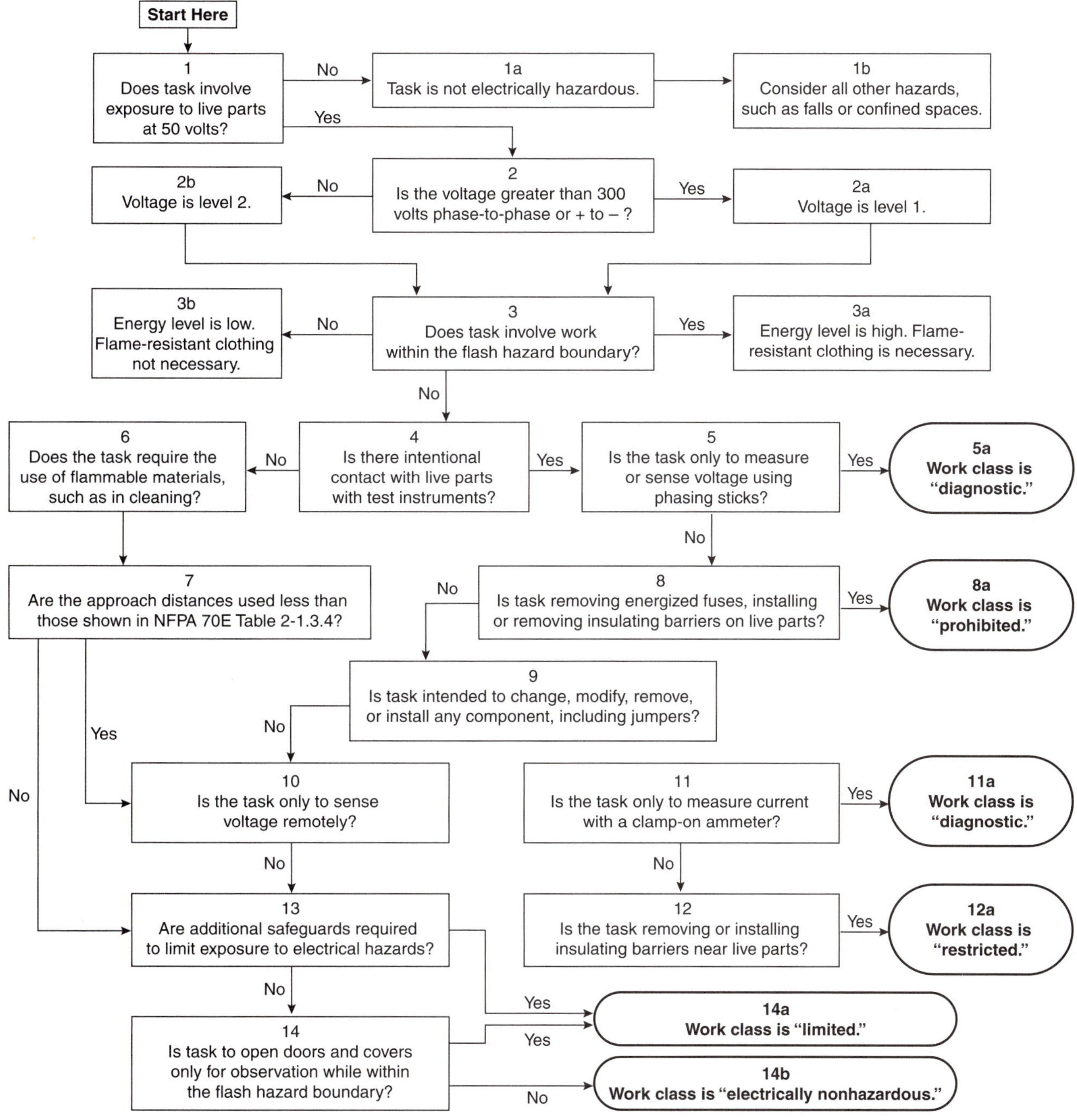

FIGURE 7-3 *Hazard Risk Analysis Flowchart*
Source: Modified from a flowchart in *Electrical Safety in the Workplace,* by Jones and Jones.

again, appropriate authorization and/or flame-resistant clothing can be selected. Proceed to block 4.

Blocks 4 and 5. In blocks 4 and 5, the objective is to begin to determine the type of task being considered. Block 4 determines if intentional contact with live parts is being planned or considered. For instance, in order to test for voltage, contact must be made. In block 5, more detail is sought about why intentional contact is anticipated. If the answer to this question is "no," the work must be diagnostic in nature. Therefore, the work category could be called "diagnostic."

Block 6. In block 6, the objective is to determine if the work task might result temporarily in an explosive environment.

Block 7. In block 7, the objective is to determine the degree of exposure to shock.

Blocks 8 and 9. In blocks 8 and 9, the objective is to continue to learn more details about the intended task. A "yes" answer suggests that repair is underway. Of course, any equipment in need of repair should be deenergized. In some instances, a repair-type task might be needed in the vicinity of a different circuit. An electrically safe working condition should be implemented prior to executing any repair work.

Blocks 10, 11, 12, and 13. The objective of these blocks is simply to gain more information about the intended task. Blocks 11a and 12 suggest that the degree of exposure to a person taking a current reading and removing barriers is unique. Appropriate category names are assigned to the work category.

Block 14. The question in Block 14 is the last to be asked. By answering "no" to all preceding questions, the task must be only to look inside the door or cover. The objective of this block is to emphasize that simply opening a door for a peek is hazardous. Energized conductors and terminals frequently exist inside the door. Of course, potential for an arc-flash event exists at any time the door or cover is less than completely secured.

When a job or task is performed routinely, it is possible to perform a hazard/risk analysis one time and then document the result of that analysis in the form of a standard operating procedure. The next time it becomes necessary to perform the task, then it will be necessary only to ensure that the hazard exposure is the same as when the analysis was performed.

When performing a hazard analysis for electrical hazards, the worker must consider the equipment that is actually installed. The overcurrent protective device might possibly have been changed since the equipment was installed. Both arc flash and arc blast depend heavily upon the clearing time of the protective device installed in the circuit, thus making the installed overcurrent protective device very important in the analysis.

■ ASSESSING RISK

Applying the principles on which the electrical safety program is based provides guidance about if and when any risk of injury is acceptable. The electrical safety program identifies who has the organizational authority to accept risk of injury.

One element of risk can be determined by comparing the contemplated task with previous experience for the same or a similar task. The length of time the worker is exposed to the hazard is a second element of risk. For example, if exposure boundaries are barely crossed, the risk of initiating an incident might be low.

A third element of risk is related to the qualification of the person(s) who will be executing the task. If the person is highly skilled, the degree of risk is less than that for a person who is only skilled. Perhaps the difference in skill level is related to the number of times the worker has executed this specific or a similar task. The skill level difference might be associated with the worker's knowledge level of electrical hazards and appropriate protective measures.

A fourth element of risk is related to the physical ability of the person who will execute the work. The worker must possess the strength and stamina to physically accomplish the task without increased exposure to the identified hazard. The worker must be able to see the complete work area directly. Adequate lighting is necessary, and the worker must be able to see the complete work area without craning his or her head into an unnatural position. Sometimes, bifocal or trifocal eyeglasses make it difficult to see the work task easily.

A fifth element of risk is related to the mental alertness of the worker(s). A worker is usually more alert early in the day. He or she likely will be able to maintain greater concentration in the morning hours. As the day progresses, workers are more likely to be distracted by other events of the day. Should the work task be planned for overtime hours, the degree of risk is elevated.

Managers generally reserve for themselves the decision associated with accepting or rejecting risk. In most instances, the plant manager, the contractor superintendent, or the owner will accept the risk only when it is slight. From a legal point of view, the highest-level person in the organization can be held accountable for any injury. That high-level manager usually wants to make certain that accepting the risk of injury is necessary.

■ EVALUATING THE RESULT OF THE ANALYSIS

When the hazard/risk analysis is complete, the task will have been assigned a work category, along with a sense of risk associated with executing the work as planned. The work categories may vary from one location to another or from one employer to another. The analysis considered in Figure 7-3 selects work categories essentially based on approach distances. The following work categories should be assigned:

- Electrically nonhazardous
- Limited
- Restricted
- Prohibited
- Diagnostic

Electrically Nonhazardous

If the work category is electrically nonhazardous, no risk of personal injury is possible from any electrical hazard. However, should the work task involve changing a circuit that controls a chemically hazardous process, the risk of upsetting the operating process may be elevated.

Limited

If the work category is limited, the requirements defined by NFPA 70E, *Standard for Electrical Safety Requirements for Employee Workplaces,* for work within the limited approach boundary apply (see Table 6-1, Approach Boundaries to Live Parts for Shock Protection in Chapter 6, Hazard Boundaries). Unqualified people must not enter the area unless escorted and supervised by a qualified person. Should the flash protection boundary extend to the limited approach boundary, all persons should be protected from the effects of any potential arcing fault.

Restricted

If a work category is restricted, the requirements defined by NFPA 70E for work within the restricted approach boundary apply (see Table 6-1, Approach Boundaries to Live Parts for Shock Protection, in Chapter 6, Hazard Boundaries). The task should be formally planned, documented, and authorized by the line organization. The worker should select and use protective clothing that is adequate to avoid potential injury from the hazard(s) identified in the analysis. The risk of initiating an incident is elevated when compared to work in the limited category. Distances are much shorter, and inadvertent or incidental movement is more likely to result in an incident. Should the flash protection boundary extend to the restricted approach boundary, all persons should be protected from the effects of any potential arcing fault.

Prohibited

If a work category is restricted, the requirements defined by NFPA 70E for work within the prohibited approach boundary apply (see Table 6-1, Approach Boundaries to Live Parts for Shock Protection, in Chapter 6, Hazard Boundaries). Crossing the prohibited approach boundary has the same degree of risk as work directly in contact with the energized conductor. Since the risk of initiating an incident is highest in this work category, prohibited work should be avoided if at all possible.

However, circumstances might exist where the high degree of risk is necessary. Should this work be necessary, the work must be guided by a written plan that is authorized by the line organization (see Appendix E, Energized Electrical Work Permit). The worker must be qualified to perform the task and specifically trained (with documentation) to execute prohibited work. Each employee must wear protective equipment that is adequate to avoid injury from any hazard identified in the hazard/risk analysis.

In such work, the flash protection boundary probably extends beyond the prohibited approach boundary. The chance of initiating an arcing fault is significantly elevated when prohibited work is performed. All persons should be protected from the effects of any potential arcing fault.

■ REFERENCES

Jones, Ray A., and Jane G. Jones, *Electrical Safety in the Workplace.* Quincy, MA: National Fire Protection Association, 2000.

National Electrical Code® (ANSI/NFPA 70). Quincy, MA: National Fire Protection Association, 2002.

NFPA 70B, *Recommended Practice for Electrical Equipment Maintenance.* Quincy, MA: National Fire Protection Association, 1998.

NFPA 70E, *Standard for Electrical Safety Requirements for Employee Workplaces,* Quincy, MA: National Fire Protection Association, 2000.

Chapter 8

Personal Protective Equipment

Some people say that the definition of a sweater is an article of clothing that a mother makes her child wear when the mother is cold. On the other hand, maybe the mother knows that the child will be playing outside, the weather is cool, and the wind is blowing, so she wants to protect the child from those elements. When children grow up, they probably are more aware of their surroundings and know when to wear a sweater. But when hazards are near, do people always know what to do or what to wear to protect themselves? Electrical hazards are especially difficult for workers to guard against without specialized training and procedures they can follow. A workplace that puts employees in an environment that contains electrical hazards must have a strategy to protect its workers.

■ PROTECTING THE PERSON

The primary protective strategy should be to *isolate* workers from electrical hazards by creating an electrically safe working condition (see Chapter 9, Lockout/Tagout). A second effective protective strategy is to identify *approach boundaries* (see Chapter 6, Hazard Boundaries) that isolate workers from a hazard. Some work tasks will not permit either of these strategies to be used. For instance, when the presence or absence of voltage must be verified, approaching a potentially energized and exposed conductor is necessary. Although less effective, a third protective strategy is to use personal protective equipment (PPE) to prevent injury in the event of an incident.

PPE is the final barrier between a release of energy and an injury. Should an incident occur while a worker is exposed to the hazard, the only chance to avoid an injury is in the integrity of the PPE that the worker previously selected and is wearing. If a release of energy occurs, the chance to choose PPE or body position is not available. If contact is made with an exposed energized conductor, the worker might not be able to move away. If an arc-flash event occurs, the worker has no chance to move away.

Workers are exposed to a myriad of safety hazards, including slips, trips, and falls. Workers should use PPE that will mitigate exposure to each hazard. The worker also must ensure that wearing PPE for protection from one hazard does not increase his or her exposure to another hazard.

A worker must recognize and understand the hazard before selecting the appropriate PPE. If the hazard is shock, then adequate shock protection must be provided and worn. If the hazard is arc flash, then arc-flash protection must be provided and worn. No PPE currently is available for protection from arc blast or from flying parts and pieces resulting from arc blast, although some PPE may provide minimal protection.

PPE must cover each part of the body that is exposed to an electrical hazard. For instance, if only hands are exposed to shock, then voltage-rated gloves will afford adequate shock protection. However, if other parts of the body might contact an exposed energized conductor, then that body part must have adequate protection as well. If a person's head and upper body are within the flash-protection boundary, then the person should wear flash-protective equipment for his or her head and upper body.

The completeness and integrity of the PPE is critical to injury prevention. Shock-protective equipment must be specified and purchased in accordance with a consensus standard that establishes a voltage rating. After the protective equipment is delivered, the integrity of the equipment must be verified from time to time to ensure that equipment has not been damaged or has not deteriorated from age or wear.

Like shock-protective equipment, equipment used for protection from the thermal aspects of an arcing fault should be specified and purchased in accordance with a national consensus standard. The equipment must be kept clean and in good repair and free from holes or tears in the fabric. The fabric must not be contaminated with grease, oil, or other flammable material. Fasteners must be in place and functional. No patches or other identifying labels should be added to the flash-protective equipment unless the patch or stitching is also flash/flame resistant.

◼ TYPES OF PPE

Voltmeters

A voltmeter usually is considered to be a hand tool, similar to a socket and ratchet. In reality, however, determining the presence or absence of an electrical hazard depends on the integrity and functionality of the voltmeter. The absence of voltage should be verified immediately before contacting an exposed conductor. The voltmeter indication verifies that the conductor may be touched. Therefore, the voltmeter should be considered a safety device and included in the general category of PPE that receives special attention and care. Only the highest quality devices should be purchased, and they should always meet consensus standard requirements.

Consensus standards published by both the Institute of Electrical and Electronics Engineers (IEEE) and Underwriters Laboratories (UL®) contain very important requirements that aid in avoiding voltmeter-related incidents. Many incidents occur with voltmeters that are set on an incorrect scale. Many others occur when the leads slip from their plugs on the meter surface. Still others occur when an internal component failure initiates an arcing fault inside the device. In some instances, voltmeter leads have a voltage rating less than the maximum scale voltage. Specifying only those meters that meet national consensus standard requirements could eliminate most of these incidents.

Solenoid-type voltage indicators function by a magnetic circuit. The solenoid armature moves in relation to the amount of applied voltage. A higher voltage causes greater movement of the armature. The magnetic circuit is not completed, which results in heating. After a while the heat generated by the incomplete magnetic circuit will raise the temperature of the conductor in the armature's electrical circuit. The amount of heat generated depends on both the voltage and the length of time that contact with the conductor exists. Solenoid-type devices are assigned a duty cycle. The duty cycle is included on the manufacturer's label. Generally, however, the duty cycle is 15 seconds. Contact with an energized conductor must not exceed the assigned duty cycle. Otherwise, the electrical insulation on the armature conductor might fail and result in a significant arcing fault.

Like all personal protective equipment, all voltmeters have limitations. It is imperative that the user understand all limitations and boundary conditions that apply to the specific device being used. Sometimes, interpreting the meter indication can be confusing. The user must understand what the indication means with certainty. For the user to be qualified to use the voltmeter, he or she must be trained to understand both limitations and interpretation.

Voltmeters always should be considered to be safety equipment.

Head Protection

American National Standard ANSI Z-89.1, "Protective Headwear for Industrial Workers," defines criteria for equipment designed and tested to protect a person's head from impact, such as bumping against an obstruction. Protection from impact is the primary objective of the standard, but it also defines criteria for electrical shock protection as an option.

ANSI Z-89.1 does not address protection from arc flash. In most instances, electrical hard hats are plastic. Although a plastic hard hat offers reasonably good protection from electrical shock, the hats offer little, if any, protection from arc flash. The plastic material of construction easily could melt onto the wearer's head if exposed to an arcing fault or even ignite. When worn within the arc-flash boundary, hard hats and bump caps should be covered by a switchman's hood.

Eye and Face Protection

Standard ANSI Z 87.1, "Practice for Occupational and Educational Eye and Face Protection," covers protection for the eyes and face. However, understanding the scope of the standard is very important. For many years, injury from projectiles has been considered a significant hazard. Equipment constructed to meet the requirements of this standard is intended to afford protection from this type of injury. The standard is not intended to provide protection from any other hazard.

The arc-flash hazard is not addressed by this standard. Equipment that is stamped or marked as meeting ANSI Z 87.1 *does not* provide necessary protection from thermal energy associated with this hazard. Protective equipment that *does* provide necessary thermal protection probably also provides protection from impact; however, the reverse is not true.

As discussed in Chapter 7, Hazard/Risk Analysis, an arc flash releases a significant amount of thermal energy. However, the electrical energy is converted to frequencies that cover the entire electromagnetic spectrum: infrared, visible, ultraviolet, etc. Eye- and face-protective equipment should afford protection from the thermal energy, the ultraviolet energy, and from impact. Eyeglasses that meet the requirements of ANSI Z 87.1 offer protection from impact and also attenuate the ultraviolet energy. Eye protection should always be worn where exposure to electrical hazards is possible.

Face shields and viewing windows of switching hoods should meet the impact criteria of ANSI Z 87.1. However, these devices must also be capable of providing protection from the thermal energy and ultraviolet energy transmitted by an arcing fault. It is not necessary that the protective equipment survive the exposure; however, it must survive long enough to provide the necessary protection.

The American Society of Testing and Materials (ASTM) identifies a test that is intended to determine the protective thermal characteristics of materials that are exposed to momentary electrical arcs. The test procedure in ASTM 1959, "Standard Test Method for Determining the Arc Thermal Performance Value of Materials for Clothing," may be used to test face shields and viewing windows. The degree of protection afforded by the protective equipment can be categorized in the same way as protective apparel, i.e., an arc-thermal protective value (ATPV) rating based on the test results. Face shields and viewing windows should provide the necessary thermally insulating qualities to protect the face from arc-flash exposure. The face shield should provide protection from the intense ultraviolet energy that is radiated from an arcing fault.

Visible light is a relatively narrow band in the spectrum. Infrared and ultraviolet energy (sometimes referred to as "ultraviolet light") are adjacent to the visible light spectrum. In most instances, a material that filters ultraviolet or infrared energy also decreases the amount of visible light that is permitted to pass. A worker's visual acuity is critical to his or her ability to avoid crossing the prohibited approach boundary or dropping a tool into an exposed energized conductor. If the face shield or viewing window limits the ability of the worker to clearly see the work area, additional lighting should be installed. The choice to continue with the work task must consider the ability to see clearly. Although not technically necessary for protection from the arcing fault, safety glasses should be worn at all times, even while wearing a face shield or switchman's hood.

Any time a worker is within the arc-flash boundary, his or her face should be protected from the potential hazard.

Footwear

Safety shoes should be worn at all times when a person is on any industrial site, regardless of the existence of an electrical hazard. However, when potential for exposure to electrical shock exists, the role of footwear takes on another meaning. Only in rare instances is the potential hazard other than shock. If the footwear is intended to serve as protection from shock, the insulation integrity is important. A test is necessary to verify that the insulating quality of the footwear is sufficient to prevent the flow of dangerous current. Everyday footwear stands a very good chance of having its insulating characteristics reduced through the rigors of normal wear. It is recommended to avoid wearing such shoes for electrical shock protection.

A test can verify the continuing insulation integrity of overshoes or boots. Standard ASTM F 1117, "Standard Specification for Dielectric Overshoe Footwear," describes a protocol to determine the insulating characteristics. When they are tested, information related to the test date must be stamped on the boot. Boots can offer protection; however, they should never serve as primary protection. They should only be used in conjunction with another, more reliable protective scheme.

Sometimes, a worker might be wearing footwear that is conductive, such as carbon-impregnated rubber-soled shoes. Since a small static discharge can be problematic for sensitive electronic hardware, the conductive footwear tends to reduce the opportunity for static buildup and subsequent discharge. Conductive footwear reduces impedance to current flow, resulting in an increased shock hazard. Conductive footwear offers help in one environment but causes a problem in another.

Ground-Fault Circuit Interrupters

It is generally accepted that 6 mA of current can flow through a person's body without causing an electrical injury. The person may feel a tingling sensation but not sustain a severe shock. A device that prevents current from exceeding the 6 mA limit would prevent severe shocks and electrocutions. A ground-fault circuit interrupter (GFCI) is designed to do just that—limit the current flow to less than 6 mA.

A GFCI works by detecting leakage current that exceeds the 6 mA limit and switching off the circuit power. The device measures the amount of current flowing in the "hot" conductor and in the neutral, "grounded" conductor. Should an imbalance occur that is greater than 6 mA between these two current measurements, the GFCI assumes that the "missing" current is flowing through the person using the circuit being protected. The device then removes the source of energy by opening an internal contact. The GFCI requires a short period of time to detect the current imbalance and remove the source of energy. However, the GFCI operates quickly enough to protect against serious injury and electrocution.

GFCIs are designed and tested to fulfill the requirements of UL 943, "Ground-Fault Circuit Interrupters." This standard requires the voltage source to be removed when the current imbalance is measured to be between 5 and 7 mA. The GFCI is also required to contain a "push to test" button that verifies the mechanical operation of the GFCI. It was suggested earlier that current flow through a body is unlikely to remain at the same level. Current flow tends to increase with duration of contact. Since a GFCI is a mechanical device, time is required for it to operate. For a GFCI to prevent electrocution, the speed with which it operates is important. The GFCI's operating speed varies, indirectly, with the amount of current imbalance. When properly used, a GFCI will prevent injury.

Overcurrent Devices

As historically designed, overcurrent protection is intended to prevent or limit *equipment damage* in the event of a short circuit or an arcing fault. Although the *National Electrical Code*® (*NEC*®) has many requirements for selecting and sizing overcurrent devices, it is important to remember that the Code is a minimum equipment standard. Selecting and installing overcurrent protection to *protect people* from injury is the best way to protect both people and equipment.

The amount of time required to sense and remove the energy source is critical to minimize the effects of any potential arc flash, pressure wave, or equipment destruction. Shorter sensing, operating, and clearing times are very important to reduce the effects of the fault. Energy that is converted to heat and pressure is exponentially reduced. If the converted energy is reduced, then potential injury is reduced accordingly. Since the human body is more easily damaged than electrical equipment, reducing the potential for injury provides commensurate reduction in the potential for equipment damage.

Overcurrent protection equipment that contains a device to sense a fault, send a signal to another device to operate, and then mechanically remove the source of energy is slower than devices that sense and act simultaneously. Melting alloy devices such as fuses probably have a shorter operating time than other types of overcurrent devices, but not always.

By definition, current-limiting devices limit the let-through current. Energy that is "let-through" by the current-limiting device reduces the power released during the arcing fault. Current-limiting devices must begin to operate within the first one-quarter cycle and completely clear the fault in less than two cycles. In most cases, the fault will be cleared in less than one cycle. However, the clearing time of a specific overcurrent device must be determined by consulting the manufacturer's literature. The clearing time will vary as the amount of fault current varies.

The thermal energy released in an arcing fault is heavily dependent on the duration of the arcing current. If the fault current exceeds the ability of the overcurrent device to clear the fault, the amount of energy released in the fault will be large. If the fault current is less than the lower limit of current limitation for a current-limiting device, the device, then, can only operate in the long-time range. The fault current will flow for much longer, and the amount of energy converted to thermal energy or a pressure wave will be significantly greater.

Overcurrent protection should be viewed as a very important personal protective process.

Protective Rubber Products—Gloves, Blankets, Sleeves

Rubber products such as gloves, blankets, and sleeves provide protection from shock by adding electrical insulation between the person and an exposed energized conductor. Since insulating qualities of these products are extremely important, they must be tested frequently to ensure that necessary values remain over time. In 29 CFR 1910.137, OSHA accepts ASTM standards as the definition of acceptable testing criteria. The ASTM standards are widely accepted for this purpose. However, as with all consensus standards, it is extremely important for the user to identify and consider any limitation of the standard.

Current ASTM standards for gloves are intended only for protection from shock. However, experience has shown that the leather protectors also provide adequate protection from thermal energy. In fact, experience shows that heavy-duty gloves with long gauntlets provide adequate thermal protection for most exposures. Leather gloves are normally stitched together with cotton thread. Experience suggests that the cotton thread probably will burn away during the exposure but will hold together longer than the arc-flash event. The authors recommend that users seek clarification of the arc-flash protection characteristics from the specific manufacturer.

Rubber protective equipment should be selected based on the voltage of the conductor that will be approached.

Clothing

As discussed in Chapter 7, Hazard/Risk Analysis, the role of clothing as personal protective equipment is to provide a barrier from the hazard and a person's body. If the hazard/risk analysis indicates a chemical hazard, as with batteries, the clothing must provide protection from exposure to the chemicals. If the hazard is thermal, the clothing must provide thermal insulation between a person's body and the thermal energy released in an arcing fault.

Selecting clothing for protection from exposure to an arcing fault is much more complex. Thermal energy released in this type of incident increases very rapidly. The longer an arc lasts, the higher the arc temperature becomes. As the degree of exposure increases, the chance of injury from the thermal exposure also increases. Clothing must be selected that will protect the person from the elevated degree of exposure.

Normal street clothing can either melt and ignite or simply ignite. Clothing made from nylon, polyester, and similar synthetic textiles can melt and sometimes ignite. Clothing made from cotton and wool do not melt; however, they can ignite. Meltable fabrics will always melt if subjected to a temperature on the order of 700–900°F. If the material ignites, the burning clothing will be on the order of 1400°F. As the material melts, it bonds to the surface of the skin and must be surgically removed. Even if the electrical arc is quenched and any resulting flames are extinguished, the injury will continue to increase until the melted textile material is cooled.

The ignition temperature of cotton clothing is on the order of 1200 to 1400°F. The cotton material will not melt and is slightly more difficult to ignite. Heavier fabrics are less likely to ignite than lighter-weight fabrics. If the cotton clothing ignites, however, the flaming clothing burns at about 1400°F and exposes the victim's skin to the high temperature for a relatively long time.

Burn injury does not happen instantly. Tissue must be exposed to an elevated temperature for some period of time. Destruction of cells can occur with extended exposure to a very small increase above normal body temperature. However, as the exposure temperature increases, the time until injury decreases. Table 8-1 illustrates the temperature that might result in a burn injury. Note that exposure temperatures are very low when compared to the temperature within an electrical arc.

Protective clothing is expected to protect a person's skin from reaching the temperature that causes a second-degree burn when exposed to the plasma temperature of an arcing fault. The amount of thermal *energy* necessary to cause a second degree burn is generally accepted to be 1.2 calories per square centimeter. As the amount of thermal energy received by the skin increases, the skin temperature increases, and the resulting injury also increases. The role of protective clothing must be to limit the amount of thermal energy received by the skin.

TABLE 8-1 Temperature/Time for Occurrence of Burn Injury

Skin Temperature	Duration of Exposure	Damage Caused
110°F	6.0 hours	Cell breakdown begins
158°F	1.0 second	Cell destruction
176°F	0.1 second	Second-degree burn
200°F	0.1 second	Third-degree burn

Source: *Electrical Safety in the Workplace,* by Ray A. Jones and Jane G. Jones. Quincy, MA: NFPA, 2000.

ASTM 1959 identifies a test that is intended to determine the protective thermal characteristics of materials that are exposed to momentary electrical arcs. The test is intended to categorize materials by identifying a thermal performance value for the material when exposed to an arcing fault. According to ASTM 1959, clothing is assigned an ATPV that corresponds to the degree of protection afforded by the clothing. Clothing must be selected and worn that has an assigned ATPV rating at least equal to the degree of the hazard (see Chapter 6, Hazard Boundaries).

TABLE 8-2 Standards on Protective Equipment

Subject	Number and Title
Head protection	ANSI Z89.1, *Requirements for Protective Headwear for Industrial Workers,* 1997
Eye and face protection	ANSI Z87.1, *Practice for Occupational and Educational Eye and Face Protection,* 1998
Gloves	ASTM D120, *Standard Specification for Rubber Insulating Gloves,* 2002
Sleeves	ASTM D1051, *Standard Specification for Rubber Insulating Sleeves,* 2002
Gloves and sleeves	ASTM F496, *Standard Specification for In-Service Care of Insulating Gloves and Sleeves,* 2002
Leather protectors	ASTM F696, *Standard Specification for Leather Protectors for Rubber Insulating Gloves and Mittens,* 2002
Footwear	ASTM F1117, *Standard Specification for Dielectric Overshoe Footwear,* 1998
	ASTM Z41, *Standard for Personnel Protection, Protective Footwear,* 1991
Visual inspection of protective rubber products	ASTM F1236, *Standard Guide for Visual Inspection of Electrical Protective Rubber Products,* 2001
Apparel	ASTM F1506, *Standard Specification for Protective Wearing Apparel for Use by Electrical Workers When Exposed to Momentary Electric Arc and Related Thermal Hazards,* 2002
Face protective products	ASTM F2178, *Standard Test Method for Determining the Arc Flash Rating of Face Protective Products,* 2002

Note: ANSI—American National Standards Institute; ASTM—American Society for Testing and Materials.

Source: Based on NFPA 70E, *Standard for Electrical Safety Requirements for Employee Workplaces.* Quincy, MA: National Fire Protection Association, 2000.

TABLE 8-3 Standards on Other Protective Equipment

Subject	Number and Title
Safety signs and tags	ANSI Z535, *Series of Standards for Safety Signs and Tags,* 2002
Blankets	ASTM D1048 *Standard Specification for Rubber Insulation Blankets,* 1998
Covers	ASTM D1049, *Standard Specification for Rubber Covers,* 1998
Line hoses	ASTM D1050, *Standard Specification for Rubber-Insulating Line Hoses,* 1990
Line hoses and covers	ASTM F478, *Standard Specification for In-Service Care of Insulating Line Hoses and Covers,* 1999
Blankets	ASTM F479, *Standard Specification for In-Service Care of Insulating Blankets,* 1998
Fiberglass tools/ladders	ASTM F711, *Standard Specification for Fiberglass-Reinforced Plastic (FRP) Rod and Tube Used in Live Line Tools,* 1997
Plastic guards	ASTM F712, *Test Methods for Electrically Insulating Plastic Guard Equipment for Protection of Workers,* 1995
Temporary grounding	ASTM F855, *Standard Specification for Temporary Grounding Systems to Be Used on Deenergized Electric Power Lines and Equipment,* 1997
Insulated hand tools	ASTM F1505, *Specification for Insulated Hand Tools,* 1994

Note: ANSI—American National Standards Institute; ASTM—American Society for Testing and Materials.
Source: NFPA 70E, *Standard for Electrical Safety Requirements for Employee Workplaces.* Quincy, MA: National Fire Protection Association, 2000.

Determining the degree of hazard is difficult. However, only by performing a flash-hazard analysis can a worker realistically select protective clothing. Appendix J, Simplified Protective Clothing Table, contains a list of work tasks where the flash-hazard analysis has been performed. To use the table, the person must make sure that the circuit parameters identified as notes to the table are similar to the parameters of the circuit being approached.

The thermal protection afforded by protective clothing varies from one exposure to another. Protective clothing identified in Appendix J is intended to prevent injury. Since the characteristics of an arcing fault vary widely, a chance exists that a burn injury could still occur, regardless of the method of selecting the clothing. However, any injury that might occur would be greatly reduced. Appendices H, I, and J provide information about thermal protection.

■ PROTECTIVE EQUIPMENT REFERENCE TABLES

Tables 8-2 and 8-3 identify national consensus standards that define requirements for equipment that affects personal safety. These tables are helpful, but they are not complete.

■ REFERENCES

ANSI Z87.1, "Practice for Occupational and Educational Eye and Face Protection." New York: American National Standards Institute, 1998.

ANSI Z89.1, "Protective Headwear for Industrial Workers." New York: American National Standards Institute, 1997.

ASTM F1117, "Standard Specification for Dielectric Overshoe Footwear." Philadelphia, American Society for Testing and Materials, 1998.

ASTM 1959, "Standard Test Method for Determining the Arc Thermal Performance Value of Materials for Clothing." Philadelphia: American Society for Testing and Materials, 1999.

Jones, Ray A., and Jane G. Jones, *Electrical Safety in the Workplace.* Quincy, MA: National Fire Protection Association, 2000.

National Electrical Code® (ANSI/NFPA 70). Quincy, MA: National Fire Protection Association, 2002.

NFPA 70E, *Standard for Electrical Safety Requirements for Employee Workplaces.* Quincy, MA: National Fire Protection Association, 2000.

OSHA Regulations 29 CFR 1910.137, "Electrical Protective Devices." Washington, DC: Occupational Safety and Health Administration, U.S. Department of Labor.

UL 943, "Ground-Fault Circuit Interrupters." Northbrook, IL: Underwriters Laboratories, 1993.

Chapter 9

Lockout/Tagout

Why is lockout/tagout so difficult for people to understand? Lockout/tagout is actually a simple concept. Sometimes people are so consumed in thinking about all of the elements in the process—the locks, keys, tags, what should be shut down, who must be notified—that they fail to see the whole picture and fail to take the process one step at a time. The only real way to prevent exposure to electrical hazards is to remove all the sources of energy and prevent them from reappearing. With no source of electrical energy present, no electrical hazard is present. If all electrical hazards are removed, no exposure exists, and without exposure, no injury can occur. That is the objective of lockout/tagout.

Many people use the terms *lockout* and *tagout* interchangeably, or together, as if they were the same thing. Actually, the terms communicate very different ideas. *Lockout* suggests an image of a lock installed so that a switch cannot be turned back on until the lock is removed. *Tagout* suggests that only a tag is installed somewhere to warn other people that they should not operate a switch. Both users of lockout and users of tagout maintain great confidence that their selected system will function satisfactorily. Unfortunately, however, neither lockout nor tagout will prevent all injuries from hazardous energy.

■ FORMS OF ENERGY OTHER THAN ELECTRICAL

For an injury to occur, a person must experience an unexpected release of energy or other interaction with an energy source. When a leak develops in a 400-pound steam line, thermal energy is stored in the escaping matter. Mechanical energy is also present in the force of 400 pounds of pressure. When a person falls from a ladder or platform, energy is generated by the acceleration of gravity. A fall injury occurs when a person hits the ground or some other object, releasing the energy. Fall-arrest harnesses with deceleration mechanisms absorb some of the energy and reduce the degree of the fall hazard. However, the best way to avoid injury from an unexpected release or interaction with an energy source is to avoid exposure to the energy.

If mechanics shut down and discharge a 400-pound steam line before they break into the line, they avoid exposure to the energy stored in the steam. If they install a lock and a lockout device (thus avoiding the possibility of repressurizing the steam line), they can safely break the line or maintain a valve after the pressure is relieved (and the hot equipment has been allowed to cool sufficiently). In this case, the lock and lockout device prevent exposure to the pressurized steam, allowing no chance for the energy to reappear. The lockout is sufficient.

Although working at elevated levels cannot always be avoided, the deceleration mechanism of the fall-arrest harness can absorb some of the energy and reduce the degree of hazard. The fall hazard still exists, but the hazard is significantly smaller.

The idea, then, is to avoid exposure to a hazard. Where exposure cannot be completely avoided, workers must take steps to mitigate the potential for injury to the maximum extent possible, such as by wearing a fall-arrest harness with an energy-absorbing device.

ELECTRICAL ENERGY

Just as with other energy sources, for an electrical injury to occur, the electrical energy must be released in the form of a fault or through physical contact with the electrical energy. Clearly, then, avoiding contact with the energy source or avoiding faulted conditions can prevent injury. In an industrial environment, faults can occur for several reasons, including equipment failure and animals contacting energized components. However, the fact that work practices initiate most faults that cause injury is well documented.

LOCKOUT IS NOT ENOUGH

When some equipment must necessarily be kept energized, lockout is one way to mitigate exposure by eliminating unnecessary sources of electrical energy. However, lockout alone is not enough to prevent exposure.

As used by most electricians, the term *lockout* suggests that a disconnecting means (e.g., a switch handle or circuit breaker) has been operated and a lock has been installed that will avoid the possibility of operating the disconnecting means until the lock is removed. Even if the circuit involves a single device, such as a motor, and the circuit can be visually observed, installing a lockout device is not enough to avoid exposure to the electrical hazard. Disconnect switches fail to completely operate, circuit breaker poles sometimes fail to open, and mechanical components of the disconnecting device can also fail.

When the path of circuit conductors is not easily observed, a new set of circumstances is introduced. If the entire path of the circuit can be seen, the disconnecting means can be identified easily. However, if the entire path cannot be seen, such as when the lines are supported by a cable tray, the wrong disconnecting means can be locked out accidentally. Industrial installations frequently contain equipment with multiple sources of energy within a single enclosure. Installing a lock does not eliminate exposure. More steps are necessary.

CREATING AN ELECTRICALLY SAFE WORK CONDITION

To completely avoid exposure to an electrical hazard, the worker must remove all potential sources of energy and then eliminate all possibility that a hazardous voltage can reappear for the duration of the work task. Not only must the work task be safe, the person performing the task must feel safe while the work is being performed.

To accomplish these objectives, each step of the following process must be performed:

1. Determine all sources of energy by reviewing up-to-date drawings.
2. Disconnect all sources of energy by operating adequately rated disconnecting means.
3. Inspect, wherever possible, energy-isolating devices for visible breaks in the power conductors.
4. Perform a voltage test to determine the absence of voltage.
5. Install grounding devices, if determined necessary.
6. Install locks and tags.

The worker must perform the work task of establishing an electrically safe work condition. As such, the work task (establishing an electrically safe work condition) must be analyzed just as any other work task would be. The analysis must determine the degree of the hazard and what type of protective equipment is necessary for adequate protection.

1. ***Review up-to-date drawings to determine all sources of energy.*** Conductors in an electrical circuit can take unexpected routes and exist in unexpected places. Temporary wiring sometimes remains in place, even though it is no longer needed. Temporary conductors sometimes are installed to permit some equipment to remain in operation. Frequently, a generator is installed to keep lights burning while a facility is shut down for a maintenance turnaround. The temporary nature of these installations requires follow-through tasks to return the equipment to permanent status.

The practice of generating and maintaining single-line diagrams is common. Most industrial facilities have diagrammatic drawings that illustrate how the electrical energy is distributed throughout the facility. Typically, single-line diagrams are supplemented with identification tags and labels. Usually, an item of equipment is assigned a name that is generated from the circuit identification that supplies energy to the equipment. Ensuring that the equipment name shown on the drawing is the same name appearing on the equipment is critical.

Electrical systems and equipment in industrial facilities rarely stay the same for more that a few months or years. New equipment is installed, additional circuits become necessary, and existing circuits are rerouted. Temporary connections also may be installed. Updating single-line drawings is very important. Each site should maintain an up-to-date set of single-line drawings. Each file copy in this set should accurately portray how the electrical system is interconnected. Equipment and circuits should be "red-lined," as necessary, to illustrate all sources of electrical energy or, more importantly, where to de-energize each circuit. The record drawing should be updated and new prints for the file copy provided.

Commercial facilities generally rely on single-line drawings when the physical building is being constructed but, afterwards, rely on identifying labels. The labels, then, must be accurate and understandable. The label should identify the equipment being served and show which equipment is providing its own source of energy.

Both commercial and industrial facilities must provide accurate information on identifying labels. All information on each label should be verified before installation. For simple electrical installations, complete and accurate labels can serve the same purpose as a single-line diagram. A satisfactory label is illustrated in Figure 9-1.

Allowing inaccurate single-line diagrams or labels to remain in place is much worse than providing no information. Drawings and labels are expected to provide solid information. Someone's life may literally depend on having accurate information.

2. ***Disconnect all sources of energy by operating adequately rated disconnecting means.*** After all disconnecting means have been accurately located, the next task is to establish a break in all conductors that supply energy to the equipment or circuit. Operating an on-off control device is not sufficient to disconnect electrical equipment or circuit.

Motor Control Center 7 A 1

Textile Production Area 2

Fed from Substation 12 AB, Circuit 7

FIGURE 9-1 *Identification Label*

Only operating a disconnecting means that positively establishes a break in the conductors supplying energy is considered to be an energy-isolating device.

Making certain that the equipment is adequately rated to break the load current is necessary. The person must know enough about the equipment being served to determine if the device is adequately rated. Comparing the load rating on the label installed by the manufacturer can make the determination. If the device is not adequately rated, it *must not* be operated under load. The best practice is to shut down the operating process with the normal control devices before operating the disconnecting means.

In consensus standards that cover lockout or tagout, disconnecting means that isolate energy are usually called energy-isolating devices. Load rating energy-isolating devices is not necessary; however, if they are not load rated, they must be labeled "not load rated."

3. *Inspect, wherever possible, energy-isolating devices for visible breaks in the power conductors.* After an energy-isolating device has been opened, the door or cover should be opened to ensure that all phase conductors have opened. Energy-isolating devices can fail. In many cases, switches are operated only infrequently. Poles sometimes fail to open. Sometimes the handle of the energy-isolating device moves, and all the poles remain closed. A typical failure mode for a disconnect switch is for one or more poles to remain closed after the operating handle has been moved to the "off" position.

One objective of an electrically safe work condition is that the worker "feels confident" that the source of energy has been removed. Opening the door or cover and looking at the opening in the power conductors helps to develop that sense of safety.

In many instances, the physical break in the power conductors is essentially hidden from view, even with the door open or the cover removed. In these cases, the succeeding steps of the process take on greater importance. The bottom line is this: opening a door or removing a cover requires the worker to be exposed to electrical conductors that are not yet in an electrically safe work condition. Consequently, the worker must use all of the protective equipment that was identified when the hazard analysis was conducted.

4. *Perform a voltage test to determine the absence of voltage.* Before a voltage test is performed, the worker should understand that a voltage might be present. Although the objective is to verify the absence of voltage, the expectation should be that a voltage exists. If a voltage exists, the testing device must meet the minimum requirements necessary to measure the expected (possible) voltage. The testing device should meet the following minimum requirements:

- The testing device must be adequately rated.
- The testing device must *not* be capable of initiating a high-energy fault.
- The testing device must be easy to understand or interpret.
- The testing device must be set to the correct scale. (Only single-function meters cannot be improperly set.)
- The probes must have a characteristic, such as a knurled section, to minimize the chance of finger slips.
- If removable leads are present, the plug-in portion of the leads must be shielded.

The voltage tester that is used to verify the absence of voltage should be considered to be safety equipment. In fact, a person's life frequently depends on the effectiveness of the voltmeter and its being used effectively. The device should be of high quality, and it should be protected and treated with respect. It should be kept in a case that is designed to provide physical protection. The meter should be frequently inspected for indications of damage. The voltage tester should be cared for like the "best friend" it is to the worker.

Since the door is open and the worker is, by definition, exposed to electrical conductors and circuit parts that might be energized, any protective equipment that was determined necessary by the hazard analysis must be in use. As a final step to ensure the integrity of the voltage-testing device, the worker should verify its satisfactory operation on a known source of voltage. The worker should then check for the presence of voltage on all conducting components of the equipment. After all tests have been made, the satisfactory operation of the voltage-testing device should again be verified on a known voltage source. If the device still functions normally and the absence of voltage was verified, the equipment can be considered to be de-energized.

5. *Install grounding devices, if determined necessary.* In every instance, where a bare outside overhead line is associated with the equipment being worked on, it is necessary to install grounding clusters on the conductors from the overhead line. A bare outside overhead line is subject to accidentally coming into contact with an energized line. An energized conductor could fall onto the de-energized line, or the de-energized line could fall onto an energized line. All outside overhead conductors are susceptible to a static discharge, such as lightning. An adequately rated and installed grounding cluster will dramatically reduce any potential exposure to an unexpected source of energy.

In some instances, an electrician only feels comfortable when the circuit being approached is grounded. Although the grounding cluster may have little direct impact on the de-energized condition, the grounding cluster might improve the comfort level of the worker.

Any grounding cluster that is installed on an electrical circuit must have a fault-duty rating at least equal to the available fault-current capacity at the point in the circuit where the grounding cluster will be installed. An inadequately rated grounding cluster introduces a very serious concern if it is subjected to the energy in the circuit. Magnetic forces can break the conductor or connector and whip it around like a bullwhip. The moving conductor will injure any person it hits. Although building an adequately rated grounding cluster on site is possible, it is unlikely that the device would pass the necessary integrity tests. The tests could be conducted on site, but the cost would be prohibitive. A much better option is to purchase manufactured grounding clusters and let the manufacturer rate them.

All grounding clusters that are installed must be removed when the work task is completed. Each grounding cluster should be assigned an identifier. A record should be kept of where each grounding cluster is installed, and that same record should be reviewed as the grounding clusters are removed.

Even though the absence of voltage has been verified in the previous step, an electrically safe work condition does not yet exist. If a grounding cluster is required by the installed conditions, then a chance still exists that a voltage can appear at the location where the grounding cluster is to be installed. If such a chance exists, the worker should be wearing any personal protective equipment that was identified in the hazard analysis.

6. *Install locks and tags.* After the equipment or circuit has been de-energized and tested for absence of voltage, and after any necessary grounding clusters have been installed, the next and final step is to block, by installing locks, the isolating devices to prevent them from operating. The worker must know that the electrical energy is controlled. Many standards require that the person who might be exposed to the electrical hazard be in control of the energy-isolating device(s). The real issue is that the worker is confident that the equipment or circuit cannot possibly be re-energized.

The equipment used in the lockout (locks, tags, and any other necessary equipment) must, of course, be substantial. The equipment must be able to survive in the environment in which it is used. For example, if the equipment is to be installed in a wet area, it must be able to function if it gets wet. If the energy-isolating device is installed in an area that is exposed to extremely cold temperatures, the lockout equipment must continue to function if exposed to extreme cold.

Sometimes a chain, hasp, clip, or other device must be used in conjunction with the lock. Perhaps several locks must be installed on the same device, or a cover installed on a handle and then locked into place. All the equipment necessary (lock, chain, hasp, etc.) to prevent the energy-isolating device from operating may be called a lockout device.

In addition to identifying the lockout device, the person-in-charge (see "Person-in-Charge," in "LO/TO Procedure Requirements," below) must also identify the people who would be exposed to an electrical hazard should the equipment or circuit be re-energized. A danger tag is normally used for this purpose. The danger tag should identify the workers by name and/or payroll number. The danger tag should indicate the reason for the lockout device and the date it was installed.

After the above six-step process has been completed, the equipment or circuit cannot possibly be re-energized. All electrical protective equipment may be removed, and the work task can be performed. Other hazards are still present, but no exposure to an electrical hazard from the target equipment or circuit will exist.

■ ESTABLISHING A LO/TO PROCEDURE

A written procedure is recommended to ensure that everyone who is or may be involved in the process of locking out electrical equipment has the information necessary to complete the process (see Appendix F, Sample Lockout/Tagout Procedure). If each necessary step and each person's responsibility are clearly defined in the procedure, the chance of success is much greater.

Every time the Lockout/Tagout (LO/TO) procedure is used, each step should be reviewed to make certain that it applies to the lockout being implemented. The procedure should be a tool to guide the lockout process. The person who is knowledgeable in the physical and electrical characteristics of all potential sources of electrical energy should conduct the review. If a step in the procedure is not followed, a satisfactory reason for the omission should be recognized.

The LO/TO procedure that applies to electrical energy can be a stand-alone electrical safety procedure, or it can be integrated into a site LO/TO procedure that covers all other sources of energy. However, the "hidden conductors" suggest that unique needs are associated with lockout of electrical energy, and extra care must be exercised. For instance, sneak circuits, backfeeds, equipment failures, and lightning are all possible reasons that a voltage might reappear at the work location. Whether integrated into an overall LO/TO procedure or into a stand-alone procedure, these concerns must be addressed.

■ LO/TO PROCEDURE

Every person who will be involved in the work task must be involved in the process of establishing an electrically safe work condition, of which lockout is but one step. Each worker might need to install his or her own individual lock on the isolating device. In fact, having each person install a lock(s) and maintain control of the key is highly desirable.

Assign a Person-in-Charge

Where the electrical system/circuit is complex and the worker is not familiar with the electrical system, *the most important issue is that someone who is familiar with the electrical system/circuit leads the entire process.* This "person-in-charge," then, should be known and recognized by all workers involved in the task as being responsible for the accuracy and completeness of the lockout process.

Create an Electrically Safe Work Condition

The procedure must contain a requirement that all sources of electrical energy be removed and a lockout device with a tag be installed. The procedure should contain a requirement that each time the lockout procedure is implemented, a plan is generated that is based on up-to-date drawing information (see "Creating an Electrically Safe Work Condition," at the beginning of this chapter). The plan must identify the limits or boundary of the safe work zone and require that workers have that information.

Test for Absence of Voltage

The procedure must contain a requirement to test for absence of voltage on each circuit conductor or component that is near the point of the work task. Note that although "near" is always defined the same way, different distances are in play, depending on the qualification of the worker (see Chapter 6, Hazard Boundaries). The voltage-testing requirement should apply to each conductor, and it should apply every time the worker approaches the work location. If the worker must leave the work location, such as for a lunch period or a bathroom break, absence of voltage should again be verified when he or she returns. The procedure must identify the person, by position or name, who has overall responsibility for the accuracy and completeness of the lockout.

Identify Locks and Tags

The procedure must identify the lockout devices and tags that will be used to establish an electrically safe work condition. Lockout devices that are acceptable by the procedure must be unique and easily identifiable. Any element of the lockout device must not be used for any other purpose.

Audit the Procedure

The procedure must be audited on a frequent basis. In fact, each time the lockout procedure is implemented, any problem with the implementation or application of the procedure should be noted to change the procedure at the next opportunity. The procedure should be reviewed to make sure that it could be applied to the work task. At least annually, a record of the audit should be established and the procedure revised, as necessary, to address all issues identified in the interim audits.

Describe Enforcement

The procedure must describe what the consequences will be of not following the procedure's requirements unless satisfactory reasoning exists. Although the procedure should be treated as a tool to guide the process, individual procedural requirements must be considered to apply to each lockout.

Clean Up the Area

The worker must identify and remove all temporary connections or wiring, unless the circuit design intends temporary conductors to remain. He or she must then measure the resistance to ensure that no faulted conditions exist. As a final step, all tools must be removed and the area cleaned up.

Notify Absent Workers

The procedure must detail the discrete steps that are necessary to remove locks and tags. Each person who installed a lockout device must remove his or her lockout device unless he or she is absent from the site. For that eventuality, the procedure must identify the process to ensure that the equipment or circuit is again safe to re-energize and to warn the absent worker before he or she returns to work.

■ RETURNING THE EQUIPMENT OR CIRCUIT TO NORMAL CONDITION

When the equipment or circuit is ready for release to production/operation or normal condition, workers must recognize that the electrically safe work condition no longer exists. Therefore, the following tasks must be completed:

- All people who were associated with the work task must be told about the change in the condition of the equipment.
- Any temporary signs or warning systems must be removed and returned to their proper storage area.
- Any installed grounding clusters must be removed, properly accounted for, and returned to their storage area.
- Each door or cover that was left ajar to permit the installation of or the marking of a grounding cluster must be closed or replaced to re-establish the integrity of the enclosure.

When the electrically safe work condition was established, the role of the person-in-charge was to ensure that all sources of energy were removed and under control. When the work task is finished, the person-in-charge has the following responsibilities:

- Ensure that the work task is indeed complete and that the equipment/circuit is safe to re-energize.
- Ensure that all lockout devices are removed.
- Account for all people who were associated with the work task and make certain that they remain clear of the equipment or circuit as it is again placed into service.

If the equipment or circuit must be temporarily re-energized or placed into service for any reason, the procedure's defined process of returning to service must be implemented in its entirety.

■ TRAINING

A procedure that will perfectly protect personnel from injury is ineffective unless a training program is designed and implemented that builds understanding of the procedure and each person's role in executing a lockout. Each person must be trained to be able to do the following:

- Recognize a lockout device and understand its purpose.
- Understand the consequences of violating the principal requirements of the procedure.
- Understand the role of the person-in-charge.
- Understand the purpose of the procedure and the purpose of an isolating device.

If an employee's job assignment changes or the equipment is modified to make installing lockout devices different in any way, the employee must be retrained to know how to control the electrical energy.

The site must maintain a record of the training (who, when, and how). An employer should document the lockout training in the employee's personnel file. At least annually, each employee should be provided with refresher training. The site must use the same process as the initial training to document refresher training or retraining.

The training program for lockout can be integrated into an overall program that trains workers to understand electrical hazards. It can also be integrated into a training effort for the overall lockout program (see Chapter 10, Training).

■ TAGOUT

Consensus standards and regulations permit the use of tagout as the primary means of controlling employee exposure to electrical energy. The authors do not recommend tagout for an energy control program except in the very rigidly controlled environment discussed as the "tags-plus" program in place within the utility industry. In all other instances, installing lockout devices on energy-isolating devices is the best way to control employee exposure to electrical energy.

Transmission Lines

Transmission circuits are basically different from other kinds of electrical systems. These circuits are intended to transport electrical energy from its source of generation to a location where the energy can be distributed to equipment that will convert the electrical energy to a more directly usable form. Since the energy-isolating device and the work location may be miles apart, expecting that the workers can visually inspect the isolating device and install personal lockout devices is impractical. Therefore, in the case of transmission circuits, needs of an energy-control program are different. The system should provide assurance to the workers that the energy is under control as well as provide a realistic means of controlling the energy. Workers must be able to trust the energy-control system.

"Tags-Plus" System

Most organizations involved in the transmission of electrical energy use an energy control system called "tags-plus." Lockout devices are not used in this system of energy control. The "dispatcher" in the operation center is normally the person-in-charge. The dispatcher operates energy-isolating devices and places danger tags on a requested circuit. After the energy-isolating device has been operated, the dispatcher communicates the operation to the crew that requested the operation. The work crew then proceeds to verify the absence of voltage at the work point by testing for voltage. When the absence of voltage has been verified, the work crew then proceeds to install grounding devices in every instance. The grounding devices are installed to establish a zone of equipotential prior to starting the work.

Training is a very critical element of the "tags-plus" program. The training must establish the understanding that a person's life depends on respecting the tagout program. The training must emphasize the enforcement mechanics for violating the "tags-plus" procedure. The authors recommend the "tags-plus" program only for utility transmission circuits and equipment for transmission lines. In all other instances, lockout devices should be used.

■ TYPES OF ENERGY CONTROL

Three distinctly different types of lockout provide adequate assurance that the electrical energy is under control. They are individual employee control, simple lockout, and complex lockout. Employees must be trained to understand how they might be exposed to an electrical hazard and how to select the appropriate type of lockout for their work task.

Individual Employee Control

In a work task where the worker is standing directly in front of an energy-isolating device (such as checking or replacing one of more fuses), a lockout device might not really be needed. If the energy-isolating device is within arm's reach of the worker and in his or her continuous, direct line of vision, and if there is a single source of energy, the addition of a lockout device will not decrease potential exposure. The work task can be executed after the worker de-energizes the circuit or equipment and determines that no voltage exists. The individual worker serves as the person-in-charge. However, if the task requires the worker to be looking in a direction away from the energy-isolating device so that it could be operated without the worker's knowledge, individual employee control must not be used.

Simple Lockout

If the work task involves a single source of electrical energy, a single set of conductors, and a single crew of workers, and if the task can be completed within one shift, a lockout device is necessary to prevent the incidental operation of the energy-isolating device. However, it is necessary that an electrically safe work condition be established before the task is performed. Although the work task may be complicated, controlling exposure to the electrical energy is relatively simple. The members of the work crew can communicate simply with one another if they are working in the same area. This lockout is considered to be a simple lockout, and a written plan is not necessary. However, it is necessary that a single person be appointed as the person-in-charge.

Complex Lockout

Work tasks that involve multiple sources of energy, multiple work crews, multiple crafts, multiple locations, multiple disconnecting means, specific shutdown sequences, or work tasks that continue for more that one shift are considered to be complex. In each instance, a person-in-charge must be appointed. All persons associated with the work task or tasks must function under the guidance of the lockout person-in-charge. The person-in-charge must ensure that an electrically safe work condition has been established before any worker is released to begin a work task that could result in exposure to an electrical hazard.

One of the most important factors in a complex lockout is the process through which each person involved in the work communicates with the others. The person-in-charge must ensure that all workers are aware of any change in the status of the lockout.

A unique lockout plan must be written for each complex lockout before the lockout is begun. An up-to-date single-line diagram should be the prime source of information that serves as the basis to generate the lockout plan. The plan must define how to account for all personnel who will be involved in the work during the lockout. The lockout plan must be reviewed with each employee who might be exposed to an electrical hazard while performing his or her task(s) and account for all concerns identified by any member of the work force.

The person-in-charge should implement the lockout as defined in the lockout plan. Any condition that is found to differ from expectations should be considered sufficient reason to stop the lockout until the lockout plan is modified, as necessary, and all workers informed of the change.

At the conclusion of work associated with a complex lockout, the person-in-charge must account for all workers who were involved in work associated with the complex lockout. Each worker must be informed that the electrically safe work condition no longer exists.

■ WHERE LO/TO APPLIES

Consensus standards and regulations all address LO/TO within the limits of their specific scopes. In 29 CFR 1910.269 (d), for instance, OSHA addresses generation, transmission and distribution. In 29 CFR 1910.333, OSHA addresses general industry, and in 29 CFR 1910.147, OSHA address all forms of energy *except* electrical energy. Each of these OSHA references attempts to cover control of hazardous energy within the scope of each appropriate section. Essential similarities exist, but the scope of each section differs. To ensure that all OSHA requirements are met, identifying the essential requirements from each section and melding them into an effective procedure are necessary. The sample procedure contained in this book is based on requirements identified in NFPA 70E, *Standard for Electrical Safety Requirements for Employee Workplaces,* and intended to address all OSHA requirements where an electrical hazard might exist.

A person injured by an electrical hazard is not influenced by whether the electrical energy source is temporary or permanent. Contact with a conductor energized at 480 volts AC has the same result, regardless if the conductor is a utilization conductor or a distribution conductor. The injury is no different if the worker is a contractor or a temporary worker or if he or she is an in-house worker. The need for lockout remains the same. LO/TO applies in all instances where the possibility of exposure to an electrical hazard is present.

Although the hazard and potential injury may be different where the source of energy is not electrical, the potential for injury exists in all instances where a work task requires a person to interact with a source of energy. LO/TO is the only realistic method to mitigate exposure to injury. LO/TO applies to all instances where a worker may be exposed to energy of any nature.

■ REFERENCES

NFPA 70E, *Standard for Electrical Requirements for Employee Workplaces.* Quincy, MA: National Fire Protection Association, 2000.

OSHA Regulations 29 CFR 1910.147, "The Control of Hazardous Energy (Lockout/Tagout)." Washington, DC: Occupational Safety and Health Administration, U.S. Department of Labor.

OSHA Regulations 29 CFR 1910.269 (d), "Electric Power Generation, Transmission, and Distribution." Washington, DC: Occupational Safety and Health Administration, U.S. Department of Labor.

OSHA Regulations 29 CFR 1910.333, "Selection and Use of Work Practices." Washington, DC: Occupational Safety and Health Administration, U.S. Department of Labor.

Chapter 10

Training

Employees of many industrial companies often have a negative attitude toward training. They might think "I have to go sit through an all-day training session—I already know all of that stuff! What a waste of time." In fact, the attitude of management also might be somewhat negative. Managers might not feel sure of their own technical expertise, yet they are reluctant to seek outside assistance. Funds for training sometimes come from the bottom of the budget barrel.

The line organization must find a way to overcome these negative attitudes and make electrical safety training a very positive experience. One of the most important planning steps is to seek input from experienced workers about subject matter and procedure content. The workers must become "owners" of the training and feel proud that their company sponsors quality training.

In the case of electrical safety, training becomes imperative for three special reasons:

- Workers must receive information that can save their lives and those of their fellow workers.
- Technology is changing rapidly in the area of electrical safety as more research information becomes available.
- Funds expended on electrical safety training have proven to result in a return on investment of between four and eight times the expenditure.

If the training is interesting, engaging, and helps the employees become proficient at safe work practices, then training can become a source of pride for the entire company.

■ EFFECTIVE ELECTRICAL SAFETY TRAINING

An effective electrical safety program depends on many different factors. Ensuring that employees understand the intricacies of the program is the primary objective of any training session. Management must make sure that the electrical safety program has procedures that enable workers to understand processes of the plant's specialized equipment and then train the workers to use those procedures. Input from the electrical workers themselves is absolutely necessary for effective procedures, as is input from the workers in the actual training programs. In addition, a training program need not be tedious and mind-numbing for the student. Because the information to be taught is so important, trainers must find techniques that will engage the students mentally.

Changing Technology

For many decades, technology has been changing. Technology associated with power and control equipment changes dramatically over time. Equipment is modified to accommodate the newer technology. Each equipment construction change affects how an installer or maintenance person will interact with the equipment.

Historically, training dollars are consumed in efforts to learn about the new technology or equipment. Vendors and manufacturers build training facilities and develop in-depth training courses to teach about the new equipment. These training efforts are important. Each worker's ability to physically perform a task depends on having the necessary skill. Usually, the manufacturer's training courses address electrical safety by historically accepted standards. The electrical safety content may be limited to protecting the equipment. In general, the only hazard considered is shock. Despite the best attempts by a manufacturer to produce safe equipment, it is the installation of that equipment that has the greatest impact on whether the installation is safe or not safe.

The Unique Nature of Electrical Hazards

The hazards of working with electrical energy are unique because they are somewhat unpredictable. To prevent an incident and resulting injury, a worker must understand electrical hazards in some detail. At a minimum, he or she must do the following:

- Know where and when an electrical hazard exists.
- Understand when and how he or she is exposed to any hazard.
- Know how to eliminate any exposure to the hazard or how to mitigate the effects of any hazard that remains while the task is being executed.
- Be familiar with and use procedures and practices that are in place on the site.
- Know and realistically accept the limits of his or her authority, knowledge, and skill.

Training should never be considered complete. Effective training requires continuing efforts. To maintain worker knowledge at an elevated level, supervisors must review safety concepts and ideas with their workers at frequent intervals.

■ ELECTRICAL SAFETY KNOWLEDGE TRAINING

Although electrical workers talked about electrical equipment "blowing up" for many years, the only generally accepted electrical hazard was shock. The thermal and blast effects of an arcing fault have been recognized only since the mid-1990s. Training programs in use at a site should include a detailed discussion of each electrical shock and arc-flash hazard.

When an instructor begins to talk about the details of electrical hazards, some students might be skeptical. Skepticism is a normal and generally healthy reaction. The training program should make a point to welcome any person who is willing to make his or her skepticism known. Swapping stories about personal experiences is one way to generate a welcoming and questioning attitude.

Although the purpose of electrical hazard training is to discuss the details of the each hazard, the objective of the training must be to build *understanding* about the hazard. If a worker understands electrical arc flash, then he or she is likely to respect the hazard and wear protection warranted by the circumstances. As an employee begins to develop a new or renewed understanding of an electrical hazards, skepticism will give way to an appreciation of the catastrophic result of an unexpected release of energy.

Safety Knowledge Training Checklist

Each employer should develop a training program that covers the content of the following points:

- The effects of current flow in human tissue
- The effects of human tissue exposure to arc flashes
- Impedance of human tissue

- Impedance of human contact
- Concept and implementation of approach boundaries
- Flashover distances at various voltages
- Concept of flash-protection boundary, including how to calculate
- Hazards associated with testing circuits
- Effective construction of safety grounds
- Hazards associated with grounding
- Effects of pressure on an enclosure
- Care and inspection of a voltmeter
- Effects of voltage on current flow
- Effects of voltage on arc flash
- Unknown electrical hazards
- Relationship of exposure to hazards and injury
- Protective characteristics of personal protective equipment (PPE)
- Construction and operation of electrical equipment
- Visual indications of an electrical hazard
- Different types of electrical hazards
- Importance of communications
- Existence and content of site procedures
- Existence and content of employer policies

Source: From *Electrical Safety in the Workplace,* by Jones and Jones.

▨ ELECTRICAL SKILL TRAINING

Skill training should involve the *application* of each element of training listed in the skill-training checklist that follows. In most instances, skill training requires a combination of classroom and on-the-job training (OJT). Skill training should be based upon the role that an employee is expected to play, such as the specialized equipment on which he or she will work. In other words, training should be designed and offered to an employee that is appropriate for his or her job assignment. Technicians who are not exposed to 600-volt circuits do not need to have skills necessary to work on or near those circuits. However, it is critical that the worker understand the limit or boundary of his or her qualification.

Skill Training Checklist

Electrical safety skill training should include information related to the elements found in the following points:

- Testing for absence of voltage
- What to do if voltage is found
- Installing and removing safety grounds
- Use of temporary grounds
- Selection and use of PPE for shock protection
- Selection and use of PPE for arc flash protection
- Selection and use of procedures
- Use of effective communication
- Effective listening
- Determining the existence of an electrical hazard
- How to avoid exposure to electrical hazards
- How to establish an electrically safe work condition

- How to establish an equipotential zone
- Inspection of equipment for missing barriers and fasteners
- Determining degree of equipment degradation
- How to minimize hazard exposure if exposure is required
- How to determine that equipment can be de-energized

Source: *Electrical Safety in the Workplace,* by Jones and Jones.

■ WORK PRACTICE TRAINING

In many instances, a young person will defer to a more experienced worker. Historically, an experienced person teaches a person with less experience how to perform a work task. The person with less experience acts as an apprentice to the more experienced person. This process is both good and bad. Traditions associated with the craft are learned and continued as new workers are taught the skills necessary to perform the job. However, the tradition may include dangerous practices. Emulation and learning do not stop when the crew goes to lunch. Discussions during lunch continue to influence the less experienced worker. The lunchtime chatter is more likely to be about previous jobs. The talk will probably include how the "old hand" took a short cut and "saved the day" by completing an assigned task rapidly.

Equipment and technology both change in a relatively short time. Work practices and the procedures that guide the practices must also change. Procedures that define the steps necessary to perform a work task should be in writing and available to the work force. These procedures are a fertile source of information that should serve as training material for workers. It is not necessary to hold a meeting with graphics and handouts in the traditional sense. Instead, procedure training can and should occur each day during the job line-up. If a major change in the site procedures occurs, a more formal setting may be required. The issue is that work practice training should be an ongoing process.

■ GOOD JUDGMENT TRAINING

Training programs should always contain examples of real-life incidents. The incident scenario should be in sufficient detail for the student to understand and visualize the environment surrounding the incident. Usually when a person makes a mistake, a lesson is learned that stays with the person for a very long time. This is especially true if a personal injury resulted from the error. If a colleague was injured or narrowly escaped injury from an incident, an employee is likely to remember that point, as well.

Real-life incidents provide a fertile source for discussion of judgment. The way that the incident is discussed is important. If colleagues and coworkers were involved, the training must avoid emphasizing weaknesses in the coworker or colleague. Instead, the discussion should emphasize judgment opportunities that could have been different. If the incident being discussed is not directly associated with employees at the site but is similar to circumstances at the site, the training could emphasize how the person(s) involved in the incident made errors in judgment.

The basic idea is that students develop experience in applying good judgment to incidents. Judgment training does not have to be formal. This training could easily be a review of an incident discussed at lunchtime or in the morning when assigning daily tasks. The incident could be one that was reported in the newspaper. (The NFPA-published book *Electrical Safety in the Workplace* contains 20 actual, real-life electrical incidents or incidents that can be used for discussion.)

INFORMATION TAILORED TO THE STUDENT

The information must be presented in a form that best suits the majority of the students. The training session should have a single, clear objective. The amount of information is less important than its quality. The information should contain data that recognizes the students' current state of knowledge. In other words, the instructor should take advantage of what the student already knows. One way to do that is to refer frequently to the student's previous experience. If the students work for a single employer, they probably will be familiar with the same circumstances. If the students work for several employers, the instructor's personal experience will become the main source of anecdotal information.

The presentation should review and discuss specific incidents that each student will understand. If the majority of the students are experienced overhead line craftspersons, the incidents that are reviewed should primarily be associated with incidents involving overhead line construction. Likewise, if the majority of the students are maintenance workers, then the instruction should select incidents that are related to maintenance activities. If the students are supervisory or management persons, the incidents should relate to systemic or process-related failures.

The learning environment is improved if the information is presented in a way that engages the student on an emotional level. Frequent use of anecdotal reference to familiar incidents that result in injury will cause the student to feel a sense of sadness. However, the presentation should also contain some humor. Melding humor and sadness into a single presentation will help the student identify with the information and keep listening. If he or she listens closely to the instructor, the student probably will retain more information. However, neither the humor nor the sadness should ever overshadow the importance of the technical information being offered.

A training program usually consists of more than one session. By combining one training session with another, more than one objective may be accomplished in the program. If more than one training session is coupled with another, the shift in information should be clear to the student. He or she must recognize when the training moves to a different topic.

If topics with multiple objectives are covered in a single training session, the student should be advised on how the subject matter is connected. The student should also be advised on how the information relates to his or her work.

USING PREVIOUSLY PREPARED TRAINING MATERIAL

The compact disc that accompanies this book contains a general electrical safety hazard training program originally created as a Microsoft® PowerPoint® presentation entitled "Electrical Safety Discussion: Personal Safety." The program can be presented in its entirety, or separate sections can be given to different groups on site, according to the training need. For instance, the section on electrical hazards should be given to every person on the site who works with electricity. All site electricians should be presented with the program in its entirety. The beginning section that discusses injuries and injury analysis should be reviewed with the management structure of the company. The program is outlined in such a way that the different sections can be presented independently or as a whole. The presentation includes notes that the presenter can easily use. The instructor might wish to use other commercially available training material and use parts of several programs to make a site-tailored training program.

Manufacturers sometimes provide training material intended to help trainees understand some aspect of specific equipment. These sources provide valuable information that may be used directly. However, the students must view the instructor as an authority. The instructor must study the material and become familiar with the details covered by the prepared program. The instructor must make the presentation his or her own.

The average student will have questions about both the information and how the information relates to his or her work assignment. The instructor must be able to provide answers to the questions. The instructor may not have the answer to every question. In that event, the instructor must admit that he or she does not know the answer but indicate willingness to locate and provide the answer in the near future.

HAZARDS TRAINING

Every electrical training program should contain at least one segment that discusses electrical hazards. Regardless of a student's experience and expertise, each training opportunity should review or provide new information about electrical hazards to which the student will or might be exposed. Greater understanding of electrical hazards and the possible consequences have a bearing on the person's behavior.

REFERENCE

Jones, Ray A., and Jane G. Jones, *Electrical Safety in the Workplace.* Quincy, MA: National Fire Protection Association, 2000.

Chapter 11

Budgeting

In a tight economy or, for that matter, in any economy, company management is always pressured to decrease the budget. Sometimes, cutting expenses also means cutting programs, especially if those programs do not directly supply revenue for the company. How can a company justify the need and expenditure for an electrical safety program? That is a good question, but perhaps a better question is this: How can a company *NOT* justify the need and expenditure for an electrical safety program?

According to the National Safety Council, "Work injuries cost Americans $125.1 billion in 1998—that's equivalent to nearly triple the combined profits reported by the top five Fortune 500 companies in 1998." And based on the NIOSH National Traumatic Occupational Fatalities (NTOF) surveillance system, electrocutions were the fifth leading cause of death from 1980 through 1992. The 5,348 deaths caused by electrocutions accounted for 7 percent of all fatalities and an average 411 deaths per year.

■ REDUCING COSTS

How can a company justify the need and expenditure for an electrical safety program? Three major categories of benefits can be attributed to an electrical safety program: moral, legal, and economic. This chapter will discuss the economics of the electrical safety program and assist those who are providing justification in the form of the budget. A safety program typically reduces the following costs for a company:

- Workers' compensation costs
- Injury costs
- Healthcare costs
- Property losses
- Insurance premiums
- Litigation costs
- Disability costs
- Business interruptions

An effective electrical safety program, then, as it reduces its injury incidence rate and corresponding costs, can be an economic benefit and provide a competitive advantage to any company.

The electrical safety program costs actually are the comparison of compliance costs versus the cost of noncompliance. The cost of insurance could be significantly affected by the recordable injuries at the company. In some instances, the reduction in these injuries provides savings that could fund the safety program. Even though the number of electrical incidents is lower than other categories, electrical injury costs are typically much higher per incident than other worker categories.

The safety budget should be considered as one of the many costs of conducting business. A typical facilities budget contains numerous items, such as rent, utilities, maintenance, and insurance. The cost of a safety program is generally an insignificant percentage of the facilities budget. Appendix K, Budget Samples and Checklists, can be used as a case study with some interesting points. The company maintains 750,000 square feet of space in three separate buildings. This type of building would have an average operating budget of $10 to $13 per square foot, if the buildings were owned, and $15 to $20 per square foot if the buildings were leased. Using these figures as a basis, the operating costs would be $7,500,000 to $9,750,000 per year if the buildings were owned.

If the building were leased, the costs would be $11,200,000 to $15,050,000 per year. As shown in Appendix K, the startup costs for an electrical safety program could be estimated at $62,000, and an established electrical safety program that a company is maintaining estimated at $34,000. From the most stringent financial analysis, those figures represent less than one percent of the operating budget. Looking at the electrical safety costs, it could be less than one OSHA fine or the cost of one electrical incident. This estimate does not even factor in the potential workers' compensation savings.

■ FORMULATING THE BUDGET OUTLINE

The first step toward developing a budget for an electrical safety program is to formulate an outline. Appendix K contains a sample budget and outline. The outline includes six major topics:

- Staff costs
- Facilities expenses
- Developmental costs
- Training
- Equipment
- Administration

Whether the company sets up the program itself or hires a consultant, the company will still need an outline to evaluate that the program properly addresses its needs, provides adequate funding, and provides a comprehensive program. The budget and the performance of the program are still ultimately the company's responsibility.

The costs of the program must be converted into benefits for expenditure justification. Reviewing the budgets of similar or new programs that have been approved at the company or the electrical safety program at another company could provide important strategic information to formulate the proposed budget. Organizational groups are also a good source for this type of strategic information. Several organizations in the facilities field have completed benchmarking studies that provide valuable information regarding safety program costs.

The initial electrical safety program budget can be very difficult to develop. All of the elements of the program must be conceptualized, and adequate funding must be provided that will be within the performance parameters of the scope and acceptable to upper management. In this analysis, some areas costs will be very clear, while others may be vague or ambiguous. Consultants and vendors are a potential source of information that can address the ambiguous areas of the budget. Initial startup and maintenance costs must be considered, as well as a further subdivision into capital and expense items. Many companies often view the capital and expense items separately for tax and asset purposes.

Staff Costs

Safety management staff costs can be significant, especially in the development of the program. Is this a newly created position or an additional assignment for an existing employee? If the staff position is a new one, the budget is fairly straightforward. From the market cost for the Safety Manager, the company should add its benefit package. Human Resources or Organizational Groups are a good resource for this information.

Even if the responsibility is assigned to an existing employee, a monetary value should be assigned to the position. By assigning this value to the position, the commitment of resources is represented that, in most cases, cannot be addressed from that person's normal assigned tasks. The company should allow the safety manager adequate time to develop and implement the plan. In addition, associated costs with this plan include resources of support staff that must assemble the safety manuals, incident tracking, record maintenance, and other support-related tasks.

Facilities Expenses

Facilities expenses are a real cost of providing the program. Does the facility contain adequate space, or must additional space be supplied? Some companies charge the individual departments a cost for their space. This cost figure must be carried as an operating expense. Does the area require remodeling? This cost figure would be a capital expense. This expense must include space and resources for the support staff. Is furniture available? Some companies have surplus furniture that can be procured for no cost. Typically, office furniture purchases are considered a capital expenditure. Purchase of a modular cubicle could cost in excess of $4,000 per station.

A computer will be required to develop and maintain the written documentation. It should contain basic office software, including Microsoft® Word®, Excel®, PowerPoint®, or similar programs, 1998 edition or later. Microsoft® Office 98®, which contains all of the above software, was used to create the forms provided with this book. Even if the company has an existing computer that can be used, computer upgrades may be necessary to address all of the requirements effectively. A larger hard drive and enhanced memory might have to be purchased to run the required programs, and additional expense to interface with the existing office software might be necessary.

The manager and the assistant should be proficient in all of the required programs. All of the forms in Appendix K provided with this book were originally set up on the Microsoft® Office® format. The written procedures and assessment forms use Word® and Excel®. The electrical hazard training program, "Electrical Safety Discussion: Personal Safety" on compact disc that accompanies this book used PowerPoint®, which is a very useful, cost-effective program for developing presentations for upper management and in-house training.

Printing capabilities also will be required. The computer should be networked to a printer for cost-effective reproduction of program books and training material for staff, contractors, and vendors. If the company already has a printer that can be used, it might need to be upgraded to accommodate the quantity and types of reproduction required to support the electrical safety program. Many companies have integrated the work-order process with the electrical safety requirements. A computer that has the capacity and capability to perform these functions can be a great asset to a company (see Chapter 12, Program Information Administration, for more detail).

Telephone and Internet service are also required. Internet service can be a valuable and cost-effective way of obtaining information and keeping current with the latest technology. Numerous search engines are available that can provide free or cost-effective information for the program development. The OSHA website is an example; the URL is www.OSHA.gov. This website contains all of the OSHA regulations for the workplace. It also contains current information on programs in progress. Another resource is the OSHA statistics and fatal facts section. This could be helpful as part of the in-house training program. This information found at the website also could be used for the electrical safety awareness program.

Reference documents are required to ensure compliance with governmental regulations and provide procedures for a company. The basic reference tools should include the Codes of Federal Regulations from OSHA. The first, 29 CFR 1910, contains Subpart S, the electrical safety requirements for general industry. In addition, OSHA 29 CFR 1926 contains Subpart K, the electrical safety requirements for construction. The complete set of OSHA regulations is downloadable, free of charge, from the OSHA website. Other valuable reference material includes NFPA 70, *National Electrical Code*® (*NEC*®), NFPA 70E, *Standard for Electrical Safety Requirements for Employee Workplaces,* and the textbook *Electrical Safety in the Work Place* by Ray and Jane Jones.

Development Costs

The development requirements may be the most time-intensive part of the process. These may include updating reference drawings, performing site and task assessments, and producing books, procedures, and training manuals. An updated set of reference drawings (single-lines) is essential. At a minimum, the up-to-date single-line diagrams are crucial to identify the sources of energy that must be controlled. Knowing the sources and controls of the energy is the only effective way to protect the employees.

A copy of these single-lines should be readily available in the facilities office as well as to the employees who work on these systems. Having a set of drawings in the maintenance area or in each building with multi-building facilities reduces the possibility of workers making assumptions about the controls of the energy sources. A facility should integrate a procedure to require that funds are allocated to update each project's drawings during the close-out phase; the facility's drawings can be brought up-to-date over a period of time.

A site assessment is required to analyze the electrical hazards at a company. Chapter 5, Site Assessment, and Appendix B, Site Assessment Checklists, provide detailed information on this subject. This chapter (Budgeting) and Appendix K, Budget Samples and Checklists, provide insight into the amount of time and resources required to perform this assessment. Once the hazards have been identified, hazard/risk analysis and task assessment are also required. Chapter 7, Hazard/Risk Analysis, contains a detailed procedure on how to perform a hazard analysis, and Appendix G, Hazard/Risk Category Selections, provides guidance on hazard/risk analysis, based on NFPA 70E.

Task assessments are very helpful in understanding the extent of the training and equipment that might be required. A complete list of task assessments, including the details of workers' requirements, is available in Appendix C, Task Assessment Checklists. Calculations also can be used to determine personal protective equipment (PPE) requirements and other equipment. The cost of software and training necessary to use task assessments also must be included in the budget.

A safety program book will be required for the workers, contractors, and vendors. The safety program itself determines whether the same book is applicable for all three categories or if different books are required. The best way to manage the safety program books is to have the master program editable to allow changes as requirements, regulations, and situations dictate. The master can be modified for the different audiences to provide applicable information. The costs of duplicating these documents and other safety-related information must also be included in the budget.

Training Expenses

Training is an essential element of the electrical safety program. The electrical safety manager (ESM) requires extensive training to understand all the electrical safety issues.

The ESM is responsible for the scope and content of the training program, which includes analyzing the level of training required for the different job tasks and for the personnel within the company. The ESM also is ultimately responsible for the in-house training. The more knowledgeable the person, the more cost effective the in-house training will be. The training venues could be a contract seminar at the facility, off-site seminars, conferences, or workshops.

How much training is required by the staff and workers and how often should the training be provided? The continuity of the workforce and workers' abilities can help to answer this question. The personnel capabilities must be evaluated and integrated into the training program. The costs include not only the training but also the lost productivity to the company.

Another often-overlooked item is the cost of contractor training. Who will provide the training, and who will pay for it? If the training is provided in-house, who pays for the lost productivity of the contractor? Incorporating safety-training requirements in the qualification of contractors could minimize these training costs. Some contractors have extensive training programs that include having all of their employees trained and becoming recipients of 10-hour OSHA Cards. If an orientation of contractors is required, the costs must be factored into the budget.

Equipment Costs

Equipment costs, including maintenance of these items, must be added to the budget. First, the ESM must determine who supplies the equipment for the workers, including in-house staff and/or contractors, uniforms, tools, lockout/tagout, testing, and other equipment. For example, the company might provide uniforms for the workers. Are these uniforms in compliance with the electrical safety program? If the uniforms are made of synthetic fibers and do not comply with the requirements of ASTM F1506, they must be upgraded. Numerous equipment items must also be considered, including head protection, eye and face protection, gloves, leather protectors, footwear, apparel, ladders, safety signs and tags, blankets, plastic guards, and insulated hand tools. Chapter 7, Hazard/Risk Analysis; Chapter 8, Personal Protective Equipment; Appendix G, Hazard/Risk Category Selections; and Appendix H, Personal Protective Equipment Matrix, provide further clarification for the company's requirements. The contractors and vendors who provide services to the company must have and use the proper equipment. Who will pay for and ensure that this equipment is adequate and is properly maintained?

DEVELOPING THE BUDGET

The flow diagram illustrated in Figure 11-1 outlines the components of a safety program and the associated costs. The flow diagram can be used to develop a budget for both an established program and a startup program.

Using the Flow Diagram to Develop a Budget for an Established Program

Block 1. Block 1a assigns staff cost for the electrical safety program. In this case the assignment is handled by existing personnel, requiring extensive time and training to develop the plan. The present salary of the maintenance manager is $70,000. Allocate 10 percent of the base salary prorated on the managers time allocation to the program. Block 1b assigns cost to the support staff and other staff who are involved with the program. Based on working conditions today, most employees have all of their time allocated to tasks. In fact, many

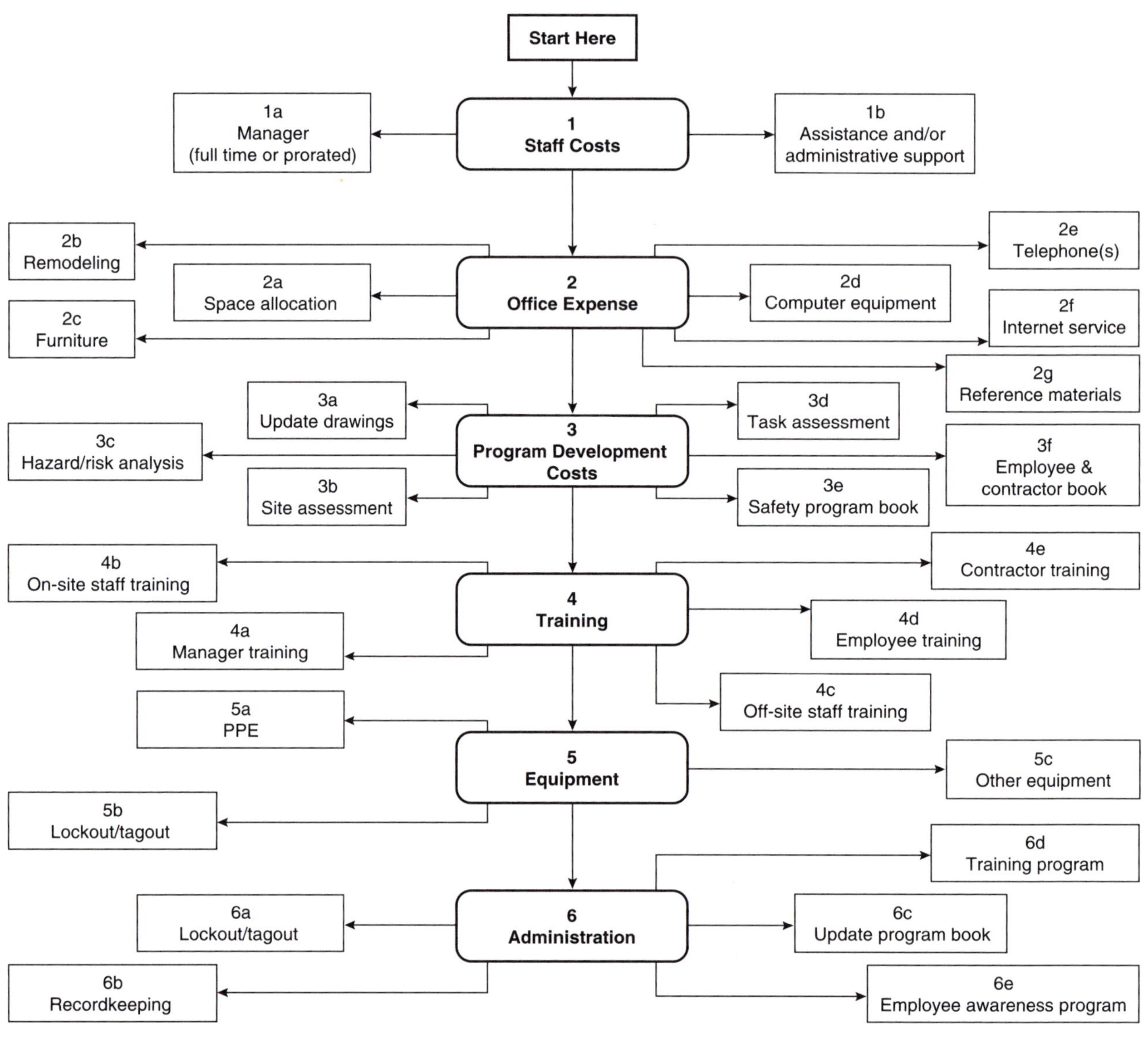

FIGURE 11-1 *Budget Flowchart*

of their tasks have to be prioritized to address the policies and responsibilities of their department. To incorporate this program would require additional time. A program of this scope would require support from the maintenance staff of 2 days per month for a total of 192 hours and support staff of 3 days per month for a total of 288 hours for updating the program book, auditing and recordkeeping, and integrating these issues into the maintenance program. As a basis for this, 192 hours for maintenance staff ($25 per hour) would equal $4,800, and 288 hours for support staff ($12 per hour) would equal $3,456. The total cost for Block 1b equals $8,256. Proceed to Block 2.

Block 2. Block 2 assigns the facilities costs. There are six items to be addressed. Block 2a is the space allocation, or additional space required. In this scenario the company uses existing space, and no costs would be assigned. If the responsibility was transferred or shared with another department, a rent allocation cost would be necessary. Block 2b assigns remodeling costs. There are no remodeling costs for the program. If there were remodeling costs, they would most likely be a capital expenditure. Block 2c is for furniture costs. Additional furniture is required. Purchasing one piece of furniture for the program would equal $500. This would be a capital expense item. Block 2d is the computer equipment cost. Block 2d assigns computer cost. The manager and support staff have computers, but they require increases in the size of the hard drive, enhanced memory, and Microsoft® Office® suite for one machine. Estimated cost is $800. Block 2e covers the telephone service. There is no additional cost because the telephones are existing. Block 2f covers Internet cost, which is vital to obtain timely information and to develop on-site training. Estimated cost is $50 per month or $600 per year. Block 2g is for reference materials. OSHA Regulations are downloadable from the Internet with no cost except for copying. The *National Electrical Code* is $55, NFPA 70E is $31, and *Electrical Safety in the Workplace* is $100. The costs are budgeted at $200. Proceed to Block 3.

Block 3. Block 3 assigns the program development costs. Block 3a assigns costs for updating reference drawings (see Chapter 11 for details). The single-lines must be field verified and updated. All other drawings must be updated on a priority basis. Time allocation of 40 hours for maintenance staff (at $25 per hour) would equal $1,000, and 40 hours for CAD time (at $25 per hour) would equal $1,000. Updating drawings would be budgeted at $2,000. Block 3b is for site assessment costs. The site assessment would be performed to ensure that all hazards have been included in the program (see Chapter 5 and Appendix B for details). Field verification of the facility is required to classify all of the work tasks. Time allocation of 40 hours for maintenance staff (at $25 per hour) would equal $1,000. Block 3c is for hazard/risk analysis costs. It is critical to understand the level of dangers that the workers are subjected to (see Chapter 7 for details). This could be done by staff if they have the proper training, or by a consultant. Assuming the staff is qualified, allocate 40 hours (at $25 per hour) to equal $1,000. Block 3d is for task assessment costs. Several methods can be used to determine task assessments. As part of the hazard risk analysis, the required information must be obtained. Another method is to use the task assessment checklist in Appendix C. These checklist items are also on the compact disc that comes with *The Electrical Safety Program Book.* Allocate 16 hours (at $25 per hour), equaling $400, to compile this information. Block 3e is for safety program book costs. The cost of developing the book is already covered in the staff costs. The allocation of money for copy costs is covered under administration. Block 3f is for employee and contractor book costs. This is also covered under administration. Proceed to Block 4.

Block 4. Block 4 assigns training costs. Block 4a is for manager training. The manager requires one training session a year, which equals $2,000. Block 4b is for on-site training for staff, including bringing in a consultant to train staff. Carry a budget cost of $2,000. Block 4c is for off-site training for staff. Send two people per year for safety training. Two sessions at $2,000 each equals $4,000. Block 4d is for employee training, including unqualified workers. Budget $500 for materials. Block 4e is for contractor and vendor training. Budget $500 for materials. Proceed to Block 5.

Block 5. Block 5 assigns equipment costs. Block 5a covers personal protective equipment. These uniforms would be covered by the maintenance budget. Any special safety equipment would either be covered by this budget or the maintenance budget. See Table 3-3.8 of NFPA 70E for items to be covered. The company would have to determine the equipment requirements and the quantities required. The sample budget has carried

$1,000, which includes voltage-rated gloves and other miscellaneous items. A PPE vendor could be a good source to assist in providing this cost. The contractors and vendors also would have to comply with the PPE requirements. Even though there may not be a cost for this item, it is important to have the contractors and vendors in compliance with the company requirements. Block 5b covers lockout/tagout equipment, which is budgeted at $200. Block 5c covers other equipment. See Table 3-4.11 of NFPA 70E for items to be covered, including items such as insulating blankets, barricades, and signage. This item would be covered in the maintenance budget. Proceed to Block 6.

Block 6. Block 6 assigns program administrative costs. Block 6a is for auditing costs. The auditing is already covered in the staff costs. Block 6b is for recordkeeping costs. This cost is already covered under staff costs. Block 6c is to update the electrical safety program book. Budget $500 to cover new technology and copying costs. Block 6d is for the training program. This includes verifying and recordkeeping of contractors and vendors and training for new staff, contractors, and vendors for $2,000. Another item is facilitation of monthly safety meetings for $1,000. Block 6e is for the employee awareness program. This could include posters, an employee awareness day, or other safety promotions for $2,000.

Using the Flow Diagram to Develop a Budget for a Start-Up Program

Block 1. Block 1a assigns staff cost for the electrical safety program. In this case the assignment is handled by existing personnel, requiring extensive time and training to develop the plan. The present salary of the maintenance manager is $70,000. Allocate 10% of the base salary for the additional responsibility and time required for this program. Block 1b assigns cost to the support staff and other staff who are involved with the program. Based on working conditions today, most employees have all of their time allocated to tasks. In fact, many of their tasks have to be prioritized to address the policies and responsibilities of their department. To incorporate this program would require additional time. A program of this scope would require support from the maintenance staff of 400 hours and support staff of 400 hours for the development of the program book, recordkeeping, and integrating these issues into the maintenance program. In the case of the costs, the maintenance and support staff work would be considered overtime. It takes a concerted effort to formulate and correlate this work into the program without distractions from the daily issues. As a basis for this, 400 hours for maintenance staff ($25 per hour at 1.5 for overtime) would equal $15,000, and 400 hours for support staff ($12 per hour at 1.5 for overtime) would equal $7,200. The total cost for 1b equals $18,000. Proceed to Block 2.

Block 2. Block 2 assigns the facilities costs. There are six items to be addressed. Block 2a is space allocation, or additional space required. In this scenario, the company uses existing space and no costs would be assigned. If the responsibility was transferred or shared with another department, a rent allocation cost would be necessary. Block 2b assigns remodeling costs. There are no costs to remodel. If there were remodeling costs, they would most likely be a capital expenditure. Block 2c assigns furniture costs. Additional furniture is required. To properly maintain the records and paperwork and store the safety related manuals, a lateral file and storage cabinet would be required. The furniture would fit into the existing space. This furniture is classified as a capital expense item. These items could be procured from existing company furniture inventory or a cost could be determined with the assistance of the purchasing department or a vendor. Block 2d is the computer equipment cost. The manager and support staff have computers, but they require increases in the size of the hard drive, enhanced memory, and Microsoft® Office® suite for one machine. Estimated cost is $800. Block 2e covers the telephone service. Telephones are existing, hence no additional cost. Block 2f covers Internet cost, which is vital to obtain timely information and to develop on-site training. Estimated cost is $50 per month or $600 per year. Block 2g is for reference materials. OSHA Regulations are downloadable from the Internet with no cost except for copying. The *National Electrical Code* is $55, NFPA 70E is $31, and *Electrical Safety in the Workplace* is $100. The costs are budgeted at $200. Proceed to Block 3.

Block 3. Block 3 assigns the program development costs. Block 3a is for updating the reference drawings (see Chapter 11 for details), with top priority being single-lines. Integrating the updating of drawings with the projects is a practical way to upgrade all of the drawings, as time permits. The single-lines must be field verified and updated. All other drawings would be updated on a priority basis. Time allocation of 80 hours for maintenance staff (at $25 per hour) would equal $2,000, and 80 hours for CAD time (at $25 per hour) would equal $2,000. Updating drawings would be budgeted at $4,000. Block 3b is for site assessment costs. The site assessment is required to build the foundation of your program (see Chapter 4 and Appendix C for details). Field verification of the facility is required to classify all of the work tasks. Time allocation of 40 hours for maintenance staff (at $25 per hour) would equal $1,000. Block 3c is for hazard/risk analysis costs. It is critical to understand the level of dangers that the workers are subjected to (see Chapter 7 for details). This could be done by staff if they have the proper training, or by a consultant. Assuming the staff is qualified, allocate 40 hours (at $25 per hour) to equal $1,000. Block 3d is for task assessment costs. Several methods can be used to determine task assessments. As part of the hazard risk analysis, the required information must be obtained. Another method is to use the task assessment checklist in Appendix C. These checklist items are also on the compact disc that comes with *The Electrical Safety Program Book*. Allocate 16 hours (at $25 per hour), equaling $400, to compile this information. Block 3e is for safety program book costs. The cost of developing the book is already covered in the staff costs. Allocate money for copy costs ($200). Block 3f is for employee and contractor book costs. They would have to be available for a number of contractors and vendors. Allocate $500 for this item. Proceed to Block 4.

Block 4. Block 4 assigns training costs. Block 4a is for manager training. The manager requires a minimum of two training sessions per year. Two sessions at $2,000 each equals $4,000. Block 4b is for on-site training for staff, including bringing in a consultant to train staff. Carry a budget cost of $2,000. Block 4c is for off-site training for staff. Send five people per year for safety training. Five sessions at $2,000 each equals $10,000. Block 4d is for employee training, including unqualified workers. Budget $500 for materials. Training is to be performed by in-house staff. Block 4e is for contractor and vendor training. Budget $500 for materials. Training is to be performed by in-house staff. Proceed to Block 5.

Block 5. Block 5 assigns equipment costs. Block 5a covers personal protective equipment. The task assessments requirements will provide a good picture of what PPE is required. If a company has noncompliant uniforms, they would have to be upgraded. These uniforms would be covered by the maintenance budget. Any special safety equipment would be covered by either this budget or the maintenance budget. See Table 3-3.8 of NFPA 70E for items to be covered. The company would have to determine the equipment requirements and the quantities required. The sample budget carries $3,000. A PPE vendor could be a good source to assist in providing this cost. The contractors and vendors also would have to comply with the PPE requirements. Even though there may not be a cost for this item, it is important to have the contractors and vendors in compliance with the company requirements. Block 5b covers lockout/tagout equipment, which is budgeted at $1,000. Block 5c covers other equipment. See Table 3-4.11 of NFPA 70E for items to be covered, including items such as insulating blankets, barricades, and signage, which is budgeted at $1,000. It could eventually be incorporated into the maintenance budget. Proceed to Block 6.

Block 6. Block 6 assigns program administrative costs. Block 6a is for auditing costs. The auditing is already covered in the staff costs. Block 6b is for recordkeeping costs. This cost is already covered under staff costs. Block 6c is to update the electrical safety program book. Budget $500 to cover new technology, copying costs, and distribution. Block 6d is for the training program. This includes verifying and recordkeeping of contractors and vendors and training for new staff, contractors, and vendors. Another item is facilitation of monthly safety meetings for $1,000. Block 6e is for the employee awareness program. This could include posters, an employee awareness day, or other safety promotions for $2,000.

Administration

Additional associated administrative costs must also be included in the budget. These costs can vary widely from company to company, depending on the complexity of the facility, the structure of the working group, and the expertise of the in-house personnel. Auditing requirements also are a part of the program. These requirements could involve in-house personnel or consultants. See Chapter 13, Auditing and Recordkeeping, for information on this subject. In addition, OSHA requires such records as injuries and training. The safety program book, procedures, and manuals should be reviewed and updated on a timely basis, which requires staff and/or consultant time and copying costs. The training program, including records, must be reviewed and updated.

An employee awareness program also could be a significant benefit for a company to showcase its program in a positive light and help to develop a safety-conscious workplace. The costs for this program might include expenses for posters, distribution of information, and, perhaps, a company safety day.

■ REFERENCES

ASTM F1506, "Standard Performance Specification for Flame Resistant Textile Materials for Wearing Apparel for Use by Electrical Workers Exposed to Momentary Electric Arc and Related Thermal Hazards." Philadelphia: American Society of Testing and Materials, 2002.

Cole, Bruce C., et al, "Creating a Continuous Improvement Environment for Electrical Safety." Paper presented at the thirty-ninth annual conference of the IAS/IEEE Petroleum and Chemical Industry Committee, San Antonio, TX, September 28–30, 1992.

Jones, Ray A., and Jane G. Jones, *Electrical Safety in the Workplace.* Quincy, MA: National Fire Protection Association, 2000.

National Electrical Code® (ANSI/NFPA 70). Quincy, MA: National Fire Protection Association, 1995.

NFPA 70E, *Standard for Electrical Safety Requirements for Employee Workplaces.* Quincy, MA: National Fire Protection Association, 2000.

OSHA Regulations 29 CFR 1900 to 1910. Washington, DC: Occupational Safety and Health Administration, U.S. Department of Labor.

OSHA Regulations 29 CFR 1926. Washington, DC: Occupational Safety and Health Administration, U.S. Department of Labor.

"Report on Injuries in America." National Safety Council, 1999 Injury Facts.

"Worker Deaths by Electrocution: A Summary of NIOSH Surveillance and Investigative Findings." Cincinnati, OH: National Institute for Occupational Safety and Health, 1998.

Chapter 12

Program Information Administration

Get ready…Get set…Go! The process of assembling all of the applicable components from the previous chapters into a comprehensive program establishes the company's vision of electrical safety. What remains is the practical administration of the program. This chapter offers some ideas on how to figure out the nuts-and-bolts administration of the program information and get started.

The overall administration of the company's electrical safety program focuses on managing the program objectives and takes into account the following factors:

- Compliance issues (see Chapter 2)
- Establishment of an electrically safe work culture (see Chapters 3 through 10)
- Administrative issues (see Chapter 11, Budgeting, and Chapter 13, Auditing and Recordkeeping)

To set up the practical administration of the program, the electrical safety manager must build the mean for efficient *dissemination of information* to all of the affected parties. The communication loop must be complete, including feedback from the workers. The program flow chart shown in Chapter 3 helps the user to understand the informational needs of a specific company.

Compliance with OSHA regulations is an important part of any electrical safety program. Because OSHA requirements are written in compliance-based language, converting them into a practical working plan for a company requires research and technical expertise. These regulations and electrical requirements must be integrated into a logical, practical, and effective program that is communicated to employees.

OSHA regulations are Federal law, but the implemention of the prescriptive and performance criteria for a safety program is the responsibility of the employer. The regulations (29 CFR 1910 and CFR 1926) tell employers what they are required to do and leave the interpretation of the regulations to the employers. In addition, OSHA regulations do not provide remediation or solutions to electrical safety issues. OSHA's general industry electrical regulations (29 CFR 1910, Subpart S) were last updated in August 1990, based on Part II of NFPA 70E, *Standard for Electrical Safety Requirements for Employee Workplaces* (1988 edition). OSHA electrical regulations for construction (29 CFR 1926, Subpart K) were last updated in 1986. Numerous enhancements in technology have occurred since these regulations were last revised. NFPA 70E parallels the OSHA documents (29 CFR 1910, Subpart S) in structure and subject matter, focusing on protecting the worker through safe work practices. NFPA 70E provides practical application to OSHA requirements. NFPA 70E is updated on a three-year cycle to be current with the latest technology. Many companies rely upon NFPA 70E to form the foundation of their program and complement it with their company's best work practices and procedures.

There are two distinct classifications for electrical safety programs. There is one program for general industry and one program for construction. The main difference is that general industry is covered by OSHA 29 CFR 1910 and construction is covered by OSHA 29 CFR 1926. The goal of both classifications of electrical safety programs is to provide a safe workplace free from electrical hazards, and at the same time be in compliance with the applicable OSHA regulations. General industry sites have a distinct logistical advantage over construction sites. General industry sites have a central location for staff and resources, the workers are familiar with the facility, and the environment and job conditions are generally stable and predictable.

As part of the process of administration of program information, the electrical safety manager will develop electrical safety books containing all of the pertinent information associated with the company's electrical safety program. Each of the program classifications will have electrical safety books in which all of the files and records are kept and that are available for all personnel to review when needed.

Figures 12-1 and 12-2 illustrate the organization of the safety program information and how it is disseminated for each classification of the electrical safety program. Each classification has a database containing the master records and forms as well as a master safety book containing redundant hard copies. The general industry program also will have a maintenance safety book as well as a service truck safety book. The construction program will have, in addition to the master book, a job site safety book and a service truck safety book

Advance preparation of these safety books can help every facility every day. A little bit of up-front, thoughtful preparation can go a long way in making every job site, whether a general industry or construction site, a safe place to work. The tools provided in these books are ready to be picked up and used.

▨ ELECTRICAL SAFETY PROGRAM INFORMATION

The safety program relies upon the efficient and timely distribution of electrical safety information to all affected parties. A set of procedures is required to control the flow of this

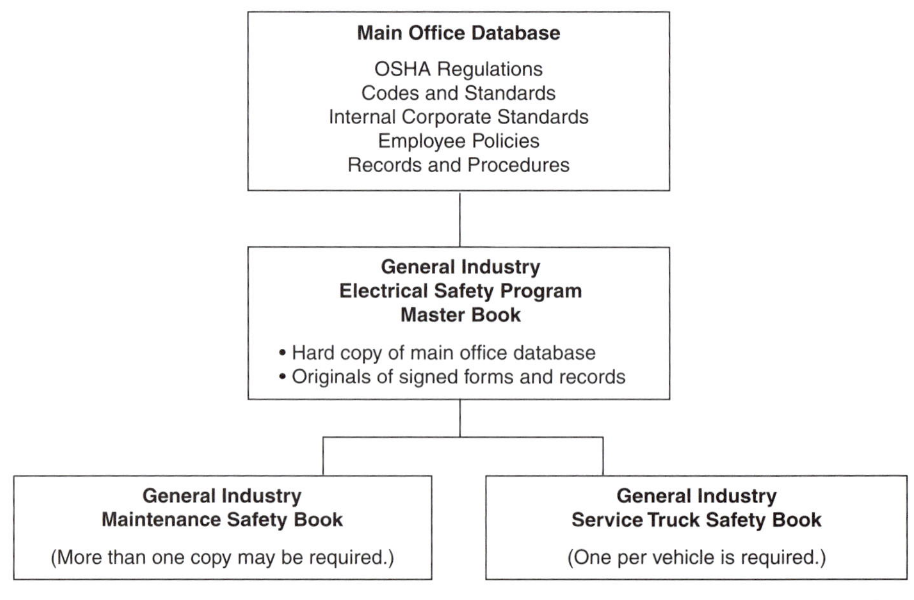

FIGURE 12-1 *General Industry Electrical Safety Program Information Flowchart*

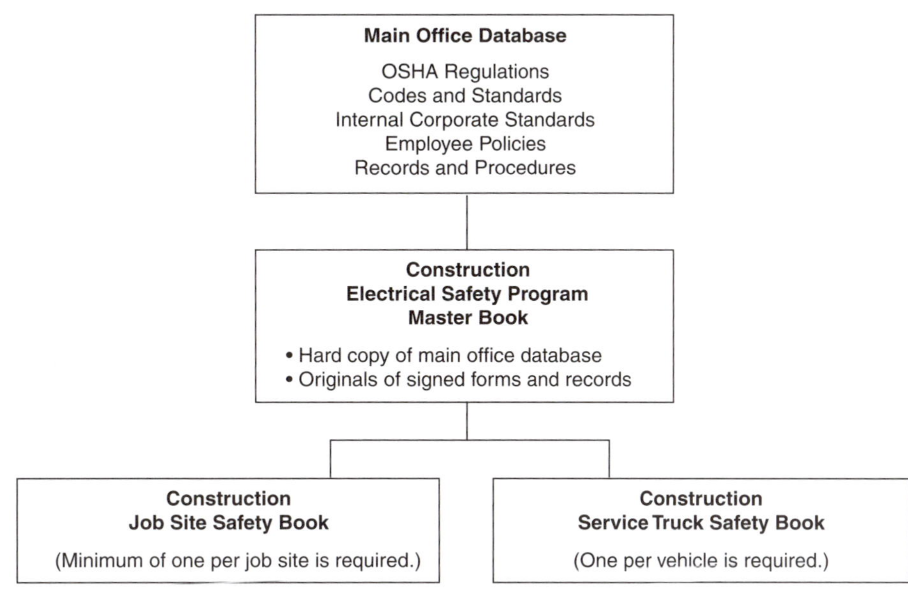

FIGURE 12-2 *Construction Electrical Safety Program Information Flowchart*

information within the company. Along with these procedures, a system must be established to compile, organize, and distribute the required information for the electrical safety program. The procedures will vary from company to company, based on the size, function, and culture of the organization. A set of standard operating procedures that can be used by all workers is required.

Maintaining records for a construction site follows the same criteria as general industry. From a functional standpoint, much of the work is similar to general industry except that the majority of the work is done at remote locations. The safety procedures must be available to a mobile workforce, including vehicles used for servicing customer needs and job sites with temporary facilities. Strong organizational skills and procedures are required to ensure that all affected workers are provided with the appropriate information and training to perform their job safely. The contractor typically does not have the same level of supervision as in general industry. Therefore, the workers do not have the same level of support and resources as in general industry.

Main Files and Documents

The electrical safety program information should always be kept in one central location. The question is, how does the company develop, store, and distribute this information? A computer system and a person competent in computer software programs (such as Microsoft® Word,® Excel,® and PowerPoint®) are the key components of cost-effective development and storage of this information.

The files and records should be set up on a database with separate folders. The first folder will be the master electrical safety program file. This folder should contain subfolders that will be the official record of the program. In addition to the computer files, a redundant hard copy—the Master Electrical Safety Book—should be maintained in the main office for both general industry and construction programs. A three-ring binder is a flexible way to maintain the hard copy of the master file, so that the information can be easily updated and readily available for reference and to copy information to assist in the safe execution of work assignments.

Maintaining a Master Set of Records

Because of the different types of electrical safety books discussed in this chapter, the main office files must reflect the different types used for all installations, whether they be general industry, construction, maintenance, or service operations. The main office should maintain a master copy of each procedure or form in all of the books that are used by all of the site workers, including those that may be for remote or mobile sites. The main office electrical safety book, or books, are the primary (master) record books, and all other books should be copied from its records.

The master set of records can be stored in the computer, a file of hard copy, or both ways. If only computer storage is used, a scanner is required to enable an accurate record of applicable data, such as field notes and signatures. A hard copy in a three-ring binder allows access to the records without requiring access to someone's computer. Both systems have advantages and disadvantages, but using both methods offers optimum flexibility. The subjects that could be part of the master set of records include the following:

- Site assessments
- Training
- Energized electrical work permits
- Auditing
- Master file for procedures
- Budgeting

Construction site records at the main office should be maintained following the same criteria as in general industry. A procedure should be established to ensure that all of the necessary paperwork is sent back to the office. All forms should be standardized to enable them to be easily tracked and recorded. The standard forms, such as Job Briefing Forms and Energized Electrical Work Permits, can be taken from Appendices D and E of this book. Reports should record on-site "tool box" or "tail gate" safety meetings. Pertinent information should be stored with the training records, including the following:

- Names of the attendees
- Subject matter
- Date and time
- A copy of the safety meeting information

Most construction projects require temporary installations. Many of the procedures, plans, and policies are similar to general industry, with distinct logistical issues. The main office function is the same as general industry. Multiple job sites and mobile equipment require the same safety information. The field books are assembled in the same manner as general industry, but distribution of this information to all of the affected workers in the company is critical. Some variables in construction, such as weather, job conditions, work on or near energized parts, and unclear direction can make implementation more difficult to monitor.

Training Records

The training records should include all of the activities of the workers, supervisors, and management (see Chapter 10). Training records are extremely valuable from a legal standpoint. Sites should keep a master log to track all of the training. The matrix shown in Figure 12-3 is an example of a method that could be used to track this information. Through the site assessment and procedures and plans processes, a list of subjects requir-

Training Subject	Name of Person Trained	Date of Training	Name of Person Who Provided Training
Electrical accident response			
Labeling, marking, and identification			
Lockout/Tagout			
Switching electrical circuits (normal and emergency)			

FIGURE 12-3 *Sample of Training Records*

ing training can be compiled. In the first column, all of the training subjects can be listed. Column 2 lists the workers requiring the specific training. Not all workers require the same training. Column 3 contains the date the worker was trained. In column 4 the names of the persons who provided the training can be listed. A log should also be maintained for contractors and vendors. The criteria for their training requirements should be in accordance with the bidder's prequalification and company policies.

Procedure Records

The procedures should be maintained in the same manner as the training records. The three-ring binders should have separators for quick access to the procedures. A procedure log should be developed to track the status and modifications of the procedures (see Figure 12-4). Column 1 lists all of the company procedures. Column 2 lists the date that the procedures were originally developed. Column 3 records the dates the procedures were modified. Column 4 contains the names of the persons responsible for the modifications. See Chapter 4 for more information on Procedures.

Audit Records

Auditing of procedures is the most effective way to confirm that the procedure performs as originally intended. (See Chapter 13, Auditing and Recordkeeping.) The audits can be

Procedure	Date Established	Date of Training	Name of Person Who Provided Training
Electrical accident response			
Labeling, marking, and identification			
Lockout/Tagout			
Switching electrical circuits (normal and emergency)			

FIGURE 12-4 *Sample of Procedure Records*

Procedure	Date Established	Audit Review Interval	Date of Last Audit	Name of Person Responsible for Audit
Electrical accident response				
Labeling, marking, and identification				
Lockout/Tagout				
Switching electrical circuits (normal and emergency)				

FIGURE 12-5 *Sample of Audit Records*

used to compare the same procedure at different locations within the company, and accurate records of all audits are essential for OSHA compliance and company requirements (see Figure 12-5). Variable circumstances at different locations can affect the effectiveness of the procedures. Column 1 lists all of the company procedures that require auditing. Column 2 should list the date that the procedure was originally developed, and Column 3 lists the audit interval requirement, which may differ from procedure to procedure, based on variables such as qualifications of personnel, risk involved in the procedure, continuity of the workforce, and the frequency that the workers perform the procedure. Column 4 contains the dates the procedures were last audited. Column 5 lists the names of the persons responsible for each audit. Maintaining a master list of procedures that are audited is an extremely useful tool. The auditing intervals can be assigned and scheduled as part of the preventive maintenance program.

■ GENERAL INDUSTRY ELECTRICAL SAFETY BOOKS

The electrical safety program for a general industry site should establish a master electrical safety book in a three-ring binder. The site electrical safety book should be divided into sections, each identified with a tab, according to the subsections that follow. The site electrical safety book should be readily accessible to the workers. A comprehensive site electrical safety book is an investment in safety and efficiency. Site workers begin to think of the book as their own personal reference when they have questions about electrical safety. Setting the book up with a table of contents and tabs for the major subject matter will make it user friendly for everyone.

General Industry Electrical Safety Program Master Book Format

Tab 1 should be the introduction. It is important to document when the program was developed and implemented and identify the key contributors. This section could also be used to convey the company commitment to this program. For example, the electrical safety program could be part of a corporate safety program. The scope, purpose, goals, or mission of the program should also be documented as a benchmark for the program. This will assist in evaluating the effectiveness and success of the program.

Tab 2 should be identified as the site assessment checklist as shown in Appendix B. It is the evaluation and record of the installation and safe work practices checklist. The title block should include the name of the building or site, or multiple buildings or sites that require additional information for clarity. It also requires the name of the person who conducted the survey and the date it was completed. This serves as a record to verify the installation is in compliance with selected codes, standards, and federal regulations and that the hazard risk assessment was completed as defined by OSHA regulations.

Tab 3 should be identified as installation issues. Site deficiencies identified by the installation checklist should be compiled and sent to the responsible party for corrective action. These deficiencies should be addressed through the maintenance work order system. Verification of the corrected conditions should be forwarded to the safety manager to update the installation checklist record.

Tab 4 should be identified as the safe work practices checklist. This checklist requires the name and date of the person conducting the survey. The information from this checklist should be separated into different groups, including procedures, hazard/risk analysis, task assessments, training, and budgeting. This information will be inserted into separate tabs.

Tab 5 is a list of necessary procedures or plans. An example is a procedure for deenergizing electrical equipment. Another procedure is required that covers lockout/tagout. A log should be developed to list all of the required procedures. This creates a quick reference to the status of what procedures are available for worker safety and when each procedure was created or last modified and by whom. The procedures can be developed using the information in Chapter 4. Having a database of these procedures makes them readily available. This tab contains the energized electrical work permit and a job briefing form. The electrical safety program should define whether or not a job briefing form (Appendix D) and/or an energized electrical work permit (Appendix E) is required.

Tab 6 is hazard/risk analysis information. The first item should be the Hazard/Risk Analysis Flowchart from Chapter 7. Next will be all of the hazard/risk assessments that were performed on site. This list of tasks that have been evaluated is an important record for audit or litigation purposes. Next should come the Hazard/Risk Category Table from Appendix G. This reference guides the thought process used in the development of the task assessments.

Tab 7 is for task assessments. Appendix C contains a detailed set of task assessments, including a title page, based on NFPA 70E. The tasks should be set up with quick reference tabs for the different equipment or task classifications. The separate sections within this tab include the following:

- 600-volt switchgear
- 600-volt motor control centers
- Panelboards rated at greater than 240 volts up to 600 volts
- Panelboards rated at 240 volts and below
- Common tasks on systems rated 600 volts and below
- Other equipment rated at 600 volts
- Metal-clad switchgear rated at 1 kV and higher
- NEMA E2 motor starters rated from 2.3 kV to 7.2 kV
- Other equipment rated at 1 kV and higher

This section facilitates a quick reference to more than 70 different work tasks that the workers might encounter. These task assessments could even be integrated into the company's work order system. For example, when assigning a work order, such as replacing a circuit breaker in an existing energized 240-volt panelboard, task assessment form B6, in Appendix C, provides the shock-protection boundary, the flash-protection boundary, and the PPE required for this task. This form could be printed out and attached to the work order. This strategy creates a distinct thought process. The worker now has all of the requirements to perform a task. He or she has a job description, location, materials required, and required safety equipment and procedures. Over a period of time, the worker becomes very familiar with the type of safety equipment required for a specific task. The ultimate goal is to have the worker consider hazards, procedures, and PPE as an integral part of every work task. These procedures and PPE are as much of a tool as a ladder, a droplight, or a drill. Depending on the work order system in place, it might be possible to develop a code that automatically selects the correct task assessment form for the work order.

Tab 8 is for selecting PPE. The electrical safety manager can insert the PPE matrix from Appendix H, which provides the detail of the PPE required based on the hazard/risk category. Next, the Protective Clothing Characteristics Table from Appendix I can be inserted. This table provides a cross reference for the hazard/risk category as compared with the calories per square centimeter of exposure. Column 1 provides the hazard/risk category, and column 2 provides the clothing description. Column 3 provides the total weight of the clothing, and column 4 provides the minimum arc-thermal protective value (ATPV) rating of PPE in calories per square centimeter. For example, a hazard/risk category 2 has a thermal exposure of 8 calories per square centimeter. When exposed to this hazard level, the worker must be provided with PPE that has an ATPV rating of at least 8 calories per square centimeter. Appendix J, the Simplified Protective Clothing Table, can be inserted next. This table provides the user with a default reference for appropriate PPE selection.

Tab 9 is for temporary installations such as power and lighting required when a company renovates or adds space to their facility. In-house personnel or contractors could perform this work. The construction checklist in Appendix B should be used to inspect the job for compliance with applicable codes and regulations. The checklist could be inserted in Tab 2 or become a separate tab of the site electrical safety book. The checklist also could be used as the basis for developing a performance specification for the work. It provides the boilerplate for installation methods and safe work practices used for the temporary work. It could also be used for periodic safety inspections of the facility.

Tab 10 is for training. The hazard risk analyses that were performed, along with the procedures that are required for the program, are the basis of the training requirements. The electrical hazard-training program that is included on the compact disc comes next in this section. The electrical hazard-training program contains training modules covering injuries, hazards, fire, electric shock, arc flash, arc blast, and responsibility. This program can be used to conduct in-house safety training for workers. The section could be broken down into smaller modules to be used as monthly safety meetings.

Tab 11 is the electrical safety program budget. The information from Appendix K is inserted here. This information includes the following:

- Budget Flow Diagram
- Using the Budget Flow Diagram
- Sample Budget Outline for an Established Program
- Sample Budget Outline for a Start-Up Program
- A Microsoft® Excel® Budget Template for an Electrical Safety Program

Using the information provided in Chapter 11 as a guide, the electrical safety team can evaluate the costs of the electrical safety program. The information in Chapter 11 and the appendix reference, Using the Budget Flow Diagram, can assist in modifying the budget costs associated with a specific program. The goal of this reference material is to have the users think about how this sample compares to his or her program and adjust the cost accordingly. In some cases the item might not be applicable.

General Industry Maintenance and Service Truck Safety Books

The General Industry Maintenance Book and Service Truck Safety Book do not require the administrative subjects that are included in the General Industry Master Book. Essentially, the Maintenance and Service Truck Safety Books contain the same information, and the most effective way to organize them is to insert all of the required material into three-ring binders to make updating and copying easier. The Maintenance Safety Book provides the required information for the central maintenance area. If the company has multiple buildings, the maintenance book should be available in each building. A copy of the Service Truck Safety Book should be kept in each service truck, making it available for reference at all field locations. In instances when supervisory personnel are not available—off hours or an emergency situation—this book can provide guidance for the worker.

General Industry Maintenance and Service Truck Safety Book Format

Tab 1 should be the introduction. It is important to document when the program was developed and implemented and identify the key contributors.

Tab 2 should contain the up-to-date site assessment checklist. This checklist helps maintenance personnel understand the magnitude of the evaluation process for a site assessment. The checklist also could be used for periodic inspections as part of the preventive maintenance program.

Tab 3 contains the work procedures. The first item should be the Energized Electrical Work Permit Form in Appendix E, followed by the Job Briefing Form in Appendix D. These two items should be readily available for the workers. As part of a standard operating procedure, these two items could be attached to the work order forms, as applicable. This tab should also contain a copy of all of the approved company procedures.

Tab 4 should be reserved for hazard/risk analysis. All of the work task assessment forms that are in Appendix C should be here, using the same format as Tab 7 of the site electrical safety book, allowing a reference by the user for a specific task. This tab should also contain the Hazard/Risk Category Table from Appendix G, providing the user with an easy lookup chart to compare different tasks. Next goes the PPE Matrix from Appendix H, which provides the PPE requirements for arc flash and arc blast. Then should come the Protective Clothing Characteristics Table from Appendix I, which provides a cross reference between the calorie exposure and the hazardous/risk classification. Over a period of time, the worker will equate the hazardous/risk category with calories, an important concept for the worker to have confidence that the appropriate level of protection has been chosen. Next should come the Simplified Protective Clothing Table (Appendix J), which provides a simplistic approach to PPE selection. At times, an employee might be called in to work from home, such as in an emergency situation or an off-hours trouble call. The worker could enter a stress-filled situation where he or she would have to restore power for the company to maintain production. Such a situation could influence the worker's normal thought process and lead to a serious incident. Having the task assessment information readily available for the worker could be the difference between life and death.

■ CONSTRUCTION ELECTRICAL SAFETY BOOKS

The administration of program information for construction sites follows the same criteria as the general industry. Included are the steps setting up a master file and procedures for distribution of safety information. The Construction Electrical Safety Master Book follows the same format as discussed for the General Industry Master Book.

There are additional considerations when assembling the construction books. If the work involves new construction, it falls under OSHA CFR 1926, Subpart K. The electrical safety manager would use the construction site assessment checklist and its requirements. If the work involved work in existing facilities, it may be under OSHA CFR 1910, Subpart S. This regulation has some requirements that are different from the construction standard. Good work practices should consider reviewing both of these regulations to provide the best protection for their workers.

Construction Electrical Safety Program Master Book Format

Tab 1 should be the introduction. It is important to document when the program was developed and implemented and identify the key contributors. This section could also be used to convey the company commitment to this program. For example, the electrical safety program could be part of a corporate safety program. The scope, purpose, goals, or mission of the program should also be documented as a benchmark for the program. This will assist in evaluating the effectiveness and success of the program.

Tab 2 is a copy of the Construction Checklist. The specific site checklists are maintained and kept at the individual sites.

Tab 3 should be identified as the temporary installation plan. One of the most beneficial projects that a contractor could undertake is to develop a temporary installation plan. The Construction Checklist in Appendix B contains the information necessary to formulate and develop a comprehensive plan, including both the installation requirements and the safe work practices that are required for the job. The installation methods can be developed in a logical order as shown in the checklist. Other techniques that typically do not exist in a general industry site can help provide for safety on a temporary site. Examples of such methods are the extensive use of ground-fault circuit interrupters or an assured grounding program. The temporary installation checklist could also be used as a site walkthrough to ensure that the equipment was not left in an unsafe condition, exposing workers to electrical hazards.

Tab 4 should be identified as necessary procedures or plans. This tab should include all of the procedures or plans for situations that could be encountered on a job site. Being proactive with these procedures is a great investment for safety and protection from liability. Some sites have unique requirements, and having this information readily available could prevent an electrical incident. A situation might arise where the company sends a worker out to a remote site that does not have a site electrical safety book. The company could copy the necessary procedure(s) to give to the worker or fax or e-mail them, if that service is available. A log, similar to the general industry log, should be developed to ensure that the procedure is safe and up-to-date.

Tab 5 should be identified as hazard/risk analysis information. Electrical hazards are the same for both general industry and construction. The same steps listed in Tab 6 for general industry should be followed.

Tab 6 should be identified as task assessments, and ***Tab 7*** should be identified as PPE selection. The same steps listed in Tab 7 and Tab 8 for general industry should be followed.

Tab 8 should be identified as training. This section should contain the electrical hazard training program from the compact disc. The electrical hazard training information, the hazard risk analyses that were performed, and the procedures required for the program form the basis for the training. The modules in the program become tools to use for worker training.

Tab 9 should be identified as the electrical safety program budget. The same steps listed in Tab 9 of general industry should be included.

Construction Job Site and Contractor Service Truck Safety Books

The Job Site Safety Book and the Service Truck Safety Book do not require all of the information that is in the Construction Master Book. A copy of the Job Site Safety Book should be kept at every job site in the job trailer or the gang box, and a copy of the Service Truck Safety Book should be kept in each vehicle. These books also should be three-ring binder formats for easy updating and copying.

Construction Job Site and Service Truck Safety Book Format

Tab 1 should be identified as the temporary installation checklist. The title block should include the name of the building or site. It also requires the name of the person who conducted the survey and the date it was completed. This checklist could be a tool used for periodic inspections as part of the preventive maintenance program. The book also provides the required information in the central maintenance area and in the service truck, making it available for reference at all field locations.

Tab 2 contains the work procedures. The first item should be the Energized Electrical Work Permit Form in Appendix E, followed by the Job Briefing Form in Appendix D. The same format and criteria as Tab 2 for the Service Truck Safety Book should then be followed.

Tab 3 should be reserved for hazard risk analysis. All of the work task assessment forms that are in Appendix C should be here, using the same format as Tab 4 of the general industry maintenance and service truck safety book format. The PPE Matrix from Appendix H should also be under the same tab, as well as the Protective Clothing Characteristics Table from Appendix I. The more familiar the worker becomes with these techniques, the better informed he or she will be.

A supply of job briefing forms and energized electrical work permits should also be included inside the front cover or some other convenient location.

■ REFERENCES

NFPA 70E, *Standard for Electrical Safety Requirements for Employee Workplaces.* Quincy, MA: National Fire Protection Association, 2000.

OSHA Regulations 29 CFR 1900 to 1910. Washington, DC: Occupational Safety and Health Administration, U.S. Department of Labor.

OSHA Regulations 29 CFR 1926. Washington, DC: Occupational Safety and Health Administration, U.S. Department of Labor.

Chapter 13

Auditing and Recordkeeping

Someone once said, "If they think something is valuable, they will measure it." Another similar thought is that if people have a good idea, or in this instance, a good procedure, they will strive to make it even better. In the area of electrical safety, a lot is at stake. Workers' lives can depend on how well a procedure works. Electrical safety procedures and practices are valuable—important parts of a safety program—and their effects should be measured, examined, and improved—*audited*. At regular intervals, each organization should conduct an audit that pays particular attention to one or two specific procedures or practices.

Once a procedure is developed, the originator typically feels a high degree of confidence that all task situations and conditions have been thoroughly addressed. He or she might radiate a pride of authorship. The originator probably feels that he or she has researched the procedure and that it will be very effective. The person might feel very comfortable with the procedure, but some of the workers might feel little confidence or anxiety about the procedure. If the originator feels challenged by the workers and becomes defensive, the workers might not express all of their concerns.

Only after a procedure is implemented can the users gain sufficient experience under different conditions to be sure that the procedure is safe and addresses all of the affected parties' concerns. Involving a cross section of affected employees in the auditing process results in the best procedure. The worker is the one using the procedure, not management. Chapter 4, Procedures and Plans, provides insight into the steps necessary to develop an effective procedure.

■ WHAT DOES AN AUDIT DO?

Auditing is the best way to evaluate the effectiveness of procedures. Electrical incidents and injuries should not be the only method of analyzing the safety of procedures. Unsafe acts and near misses are very important factors to consider when auditing procedures. The safety triangle in Chapter 3, Setting Up the Electrical Safety Program, shows the relationships of worker behaviors to injuries and fatalities. The triangle specifically shows that unsafe behaviors cause a huge number of near misses for every recordable injury. Correcting unsafe behaviors and near misses can dramatically increase the safety margin.

Site and procedure auditing must be considered a part of a continuous improvement process. The process must be conducted in such a way that employees feel their input is important and that refining the procedures is a team effort. Workers must be confident that they are a part of the team. Workers must not dread the process because they fear repercussions when deficiencies are uncovered. If the team effort is genuine and improving work practices is viewed as very positive, workers will view auditing in a positive manner. Staff meetings and safety talks provide an opportunity to discuss procedures. Being open to critiquing a procedure brings out points in the procedure that the workers might be uncomfortable with or don't clearly understand.

The auditing process must focus on enhancing workers' safety. Workers must perceive the audit of safety procedures as focusing on the procedure instead of the workers.

Interactions between workers and the audit team must be helpful to both workers and the company. Conversations with workers and among audit team members must avoid comments about workers and focus on either the procedure or the overall company. The company philosophy for auditing sets the tone for communications and teamwork. How the audit team gathers information sends a message from senior management about the importance of workers' safety.

■ SETTING UP THE AUDIT PROCESS

Improving the safety aspect of the work environment is the objective of auditing procedures. Obtaining data from the audits that can be analyzed enables the team to gain insight that can enhance that objective. Chapter 4, Procedures and Plans, offers a starting list of forty procedures that companies might want to consider. However, procedures must always be tailored to the needs of individual companies. The auditing process should be set up in the same format as a procedure. Consistent and objective steps must be followed to establish a benchmark for the procedures. When a site assessment is performed, potential procedures can be noted.

The company or site might want to develop an auditing checklist as a helpful tool (see an example in Figure 13-1). Such a checklist establishes a responsible party to oversee the procedures, indicates when the original procedure was established, and provides a detailed history of when it was audited. Of course, the audits will not take place at the same time or date, so the checklist will serve only as a record of the different dates. Updates based on new regulations or changing technology can be recorded. The checklist should be maintained with all of the safety records.

Procedure	Responsible Person	Date Procedure Established	Auditing Interval	Name of Auditor and Date Completed
Electrical safety program				
Electrical diagrams— updating and using				
Creating an electrically safe work condition				
De-energizing electrical equipment				
Developing work plans				
Lockout/tagout				
Personal protective grounds				
(add additional site procedures)				

FIGURE 13-1 *Sample of an Auditing Record Form*

In an auditing record form similar to the sample in Figure 13-1, the first column might describe the procedure. The second column could designate the person responsible for the specific procedure. In a smaller company, the safety manager might be responsible for all of the procedures. In a larger company, the responsibility for some of the procedures can be delegated to more than one person. The third column might be for the date that the original procedure was established, thus providing a benchmark that relates to the federal regulation and any other requirement that might have been used in the thought process to develop the procedure. The fourth column could provide the auditing interval. Each procedure should be considered separately based on the frequency of the task involved and the level of risk associated with the procedure. This interval column could be used similarly to a work order form for maintenance personnel to ensure that the audit takes place. The audit could become a goal in the responsible persons' performance evaluation.

How does the safety program address the contractor and vendor's procedures? Contractors' and vendors' procedures should be consistent with the company procedures to ensure safety. Who is responsible for the contractor's procedures and auditing? This issue could be identified in the prequalification of the contractors or be part of the project bidding specification.

■ CONDUCTING THE AUDIT

Depending on the size of the organization, the electrical audit team should include one electrician, one supervisor, and one safety professional. One team member should be appointed to record the results of the audit. The objective of the audit should be to determine if the electrical safety program should be upgraded or if workers adequately understand the procedure that is being targeted during the walk around the site.

Audit team members should choose a single procedure or task and then physically observe jobs or tasks being performed when they arrive at the scene. Team members should talk with the worker(s) to gain an understanding of what the worker(s) assignment is. The person elected to record the audit results should make notes while the audit is being conducted that will enable a more formal report to be generated later in the day.

If the audit uncovers a deficiency in a procedure, the team can formulate a resolution and modify the procedure. The procedure revision should involve workers that are affected by the deficiency to ensure that the resolution addresses all of the known hazards and does not create an additional hazard. After the procedure has been modified, the workers who will be affected can be trained in the new procedure.

■ AUDITING BEHAVIOR

Everyone has to deal with the facts of reality every day. In an industrial setting, engineers, managers, and workers deal with hard information and facts: exact engineering data and pure logic. In the aftermath of an incident, everyone wants to *do* something to make the situation better—to add a guard, pressure relief vents, improved overcurrent protection—*something!* This very human response is an important element in making the situation better and preventing injury "next time."

When a mechanical or hardware modification doesn't provide the solution, the next step is usually to add personal protective equipment (PPE) in an effort to avoid an injury in case of equipment failure. These adjustments could help to prevent maybe one-third of the incidents and injuries, but many of the incidents and injuries would still occur, because two-thirds of them are associated with behavior. When a company analyzes its physical audits, it should be done from a behavioral perspective.

Talking directly with employees is the most effective way to assess behavior. Such discussion should not involve a team concept but should simply be a one-on-one conversation to try to find out why people act as they do. These discussions should never involve recording a worker's name, and any "reasons" for behavior should be noted anonymously. Much information can be gained for the organization that can help modify behaviors. One way to modify behavior is to change the systems or processes *and* procedures.

■ TRAINING AND DOCUMENTATION AUDITS

Because consensus standards and OSHA require companies to keep several types of records, maintenance of these records becomes very important. The records should be audited at least once a year. One of the best ways to be sure training records are readily accessible is to maintain them in individual personnel files. Job descriptions should include a list of the training requirements a person must have to execute a job as well as a record of the training completion dates.

Employee records these days might be kept as data in a computer program that can actually schedule employees for training. Auditing of these records should not turn up any deficiencies, but the audit information for training records must still be maintained, as OSHA (and other external agencies) has a legal right to review it. A hard copy of these audit records should be maintained and should be no older than 12 months. Any deficiencies found by the audit must be dealt with effectively and a record kept of how they were corrected.

■ INVESTIGATING AN INCIDENT

If the auditing program is reactive, all of a company's audits will start with an incident investigation. Investigating an incident takes time and talent to understand the root cause and come to a solution that can correct the behavior or procedure. A qualified investigator must be selected, and then a team should follow a thorough process to reach a valid conclusion. To get to the root cause of the incident takes a keen insight into the procedure itself. If the investigator is an existing staff member, he or she should be given time away from his or her normal assignment, which could create a work continuity problem. If a consultant is required, significant unbudgeted cost might be involved.

The primary purpose of incident audit is to prevent similar incidents in the future by eliminating actions and behaviors that caused the original incident. The company should develop an Incident Investigation Form that includes the date of the incident, job location, names of those involved, etc. The following information should be included on the form:

- The scope of the investigation
- Names(s) of investigator(s)
- Tasks assigned to investigator(s)
- Description of the incident
- Normal operating procedures that should have been used
- Any relevant maps or sketches
- Confirmation of accuracy of maps and sketches
- Names of witnesses
- Event(s) that preceded the incident

- Interviews with victims and witnesses
- Notes on abnormal occurrences, when they were first noted, how they occurred
- Determination of root cause(s), including a likely sequence of events, probable causes, and possible alternative sequences
- A summary report that includes recommended actions to prevent a recurrence

All of the information must be compiled, sorted, and organized to determine if adequate information is available to develop a solution. Incident investigation can be a time-consuming task. At the same time, the person investigating the incident might be distracted by his or her daily responsibilities within the company. And then, once the cause of the incident has been determined, a solution must be formulated. This phase of the investigation could involve several meetings with other people in the company to establish that the chosen solution will correct the problem. The procedures or the operating manuals will then have to be modified, and the affected workers will have to be retrained on the new techniques.

Following an incident, an audit might uncover some issues of preparation for emergencies. Being prepared for emergency situations requires preliminary planning, and several issues should be addressed, such as the following:

- Will workers be able to call for help? Are telephones nearby?
- Is emergency information visibly posted where the information can be readily seen?
- Have discussions with nearby medical facilities taken place to discuss emergency possibilities?
- Are CPR-trained workers available on all shifts?
- How can workers know when visitors, such as utility workers or vendors, are on site?

Other questions that are unique to a specific company or site will arise once this discussion takes place. Workers in each job assignment should address their own specific needs and conditions as they audit their emergency procedures.

Appendix A
Electrical Safety Program Outline

This appendix contains a sample outline that employers can use to define their electrical safety programs. Adapted from NFPA 70E, *Standard for Electrical Safety Requirements for Employee Workplaces,* this outline primarily supplements Chapter 3, Setting Up the Electrical Safety Program, and Chapter 4, Plans and Procedures. The appendix also serves as a useful reference elsewhere in the book as specific components of the program are considered.

Each employer must understand that an effective electrical safety program (ESP) cannot be duplicated from any source, including this book, and implemented directly. Satisfying regulatory requirements might be possible by duplicating a program from an external source. However, for an electrical safety program to be effective, each element of the program must be discussed and analyzed on the basis of the physical environment and experience level of both supervisors and workers.

An electrical safety program should be communicated to workers in writing. Senior management must "sign off" on the content of the program in such a way that workers and members of the line organization become aware that senior management supports all elements of the program. The ESP can be integrated into an overall corporate or company safety program; however, the unique characteristics of electrical hazards and how workers are exposed to them must be considered in the program. The written program should contain the following components.

■ SCOPE

This section of the ESP should clearly state the following:

- If the program is mandatory
- If the program applies to all workers, including contractors, vendors, and service personnel
- If procedures may be supplemented by other documents
- Who is responsible for administering the program
- If any boundary conditions or limits exist

■ PHILOSOPHY

This section should clearly state the basic corporate or company philosophy on which the ESP is based. For instance, a philosophy might state that all injuries are preventable. This section should indicate what type of injuries are not considered to be preventable, if any. If working safely is a condition of employment, the philosophy section should so state.

PERSONAL RESPONSIBILITY

This section should clearly state to what accountabilities each worker would be held. For instance, if each worker is expected to be responsible for his or her own safety, this section should so state. If an employee is responsible for reporting unsafe conditions, this section should state that expectation as well.

ELECTRICAL SAFETY TEAM

This section should assign an electrical safety team and define how the team reports to senior management. This section should establish expectations for the electrical safety team and also define routine measurements to monitor the effectiveness of the team, if appropriate.

PROCEDURES

Written procedures are one of the most important features of a good ESP and a key part of a company's plan for electrical safety. They must always be written especially for the actual site conditions, specifically for the equipment in use at the site. Chapter 4, Procedures and Plans, discusses procedures in depth and includes a list of procedures a company might find appropriate.

When writing procedures for the corporation, company, or site, the electrical safety team should consider including the following:

- Purpose of task
- Qualifications and number of employees to be involved
- Hazardous nature and extent of task
- Limits of approach
- Safe work practices to be utilized
- Personal protective equipment involved
- Insulating materials and tools involved
- Special precautionary techniques
- Electrical diagrams
- Equipment details
- Sketches/pictures of unique features
- Reference data

CONTROLS

This section of the ESP should identify any controls that an employee is expected to observe. For instance, if no bare-hand contact is permitted, this section should indicate that. This section should address procedures, qualified employees, deenergizing, and hazard/risk analysis. Electrical safety program controls can include, but are not limited to, the following:

- Every electrical conductor or circuit part is considered energized until proven otherwise.

- No bare-hand contact is to be made with exposed energized electrical conductors or circuit parts above 50 volts to ground, unless the "bare-hand method" is properly used.
- Deenergizing an electrical conductor or circuit part and making it safe to work on is in itself a potentially hazardous task.
- The employer develops programs, including training, and the employees apply them.
- Use procedures as "tools" to identify the hazards and develop plans to eliminate/control the hazards.
- Train employees to qualify them for working in an environment influenced by the presence of electrical energy.
- Identify/categorize tasks to be performed on or near exposed energized electrical conductors and circuit parts.
- Use a logical approach to determine potential hazard of task.
- Identify and use precautions appropriate to the working environment.

PRINCIPLES

This section should identify any personal or corporate principles that are embraced or are expected to be embraced by employees. Electrical safety program principles might include, but are not limited to, the following:

- Inspect/evaluate the electrical equipment.
- Maintain the electrical equipment's insulation and enclosure integrity.
- Plan every job, and document first-time procedures.
- Deenergize, if possible.
- Anticipate unexpected events.
- Identify and minimize the hazard.
- Protect the employee from shock, burn, and blast, and other hazards that are due to the working environment.
- Use the right tools for the job.
- Assess people's abilities.
- Audit these principles.

STANDARD OF PERFORMANCE

This ESP section should identify any expected standard of performance. For instance, if all work is expected to meet the requirements of a specific consensus or regulatory standard, this section should indicate which standard(s).

TRAINING

This section should identify training that is expected to be executed by a site or organization. The section should indicate when specific training is required for some work categories and specify requirements, intervals, or conditions for retraining, including whether or not contract employees must receive the same training. Information on how training records are maintained should also be included in this section.

AUDITING

This ESP section should discuss any routine specific and special audit requirements, defining who has responsibility for content and who is to conduct audits.

◼ POLICIES

Any policies that address personal electrical safety should be defined in this section. Each employer should consider establishing a policy that covers the following issues:

- Standards—Which standards are expected to be routinely implemented on the site?
- Documentation—Maintaining diagrammatic and other essential drawings. Who is accountable for ensuring that drawings meet the needs?
- Abandoned cables—This policy should define a standard practice related to electrical cables and conduit that are no longer needed. If the abandoned equipment is to be removed or marked as no longer in use, the policy should be defined.
- Excavation—This policy should define how underground electrical cables and conduit are to be located before the excavation is begun.
- Procedures—This policy should define how the organization is expected to use corporate or site procedures. This section should identify who has responsibility for generating and maintaining site or corporate procedures, including how the procedures will be maintained.
- Enforcement—This policy should clearly define any routine enforcement actions that will be taken, if any. If enforcement is not considered to be routine, the policy should so state.

Appendix **B**

Site Assessment Checklists

This appendix contains two site assessment checklists that can be used in the development of a company's electrical safety program. As discussed in Chapter 5, Site Assessment, the General Industry Installation Checklist is useful in evaluating permanent installations for compliance with consensus standards and OSHA regulations. Owners and electrical construction or maintenance contractors can use the Construction Checklist to evaluate temporary installations and construction facilities for compliance with consensus standards and OSHA regulations.

To provide easy access to information in regulatory and consensus documents on relevent inspection and assessment activities, the checklists include cross-references to the following:

- OSHA 29 CFR 1910, Subpart S
- OSHA 29 CFR 1926, Subpart K
- NFPA 70E, *Standard for Electrical Safety Requirements for Employee Workplaces,* both the 2000 edition and the 2004 edition
- The *National Electrical Code®* (*NEC®*), 2002 edition

In addition, each checklist includes a section of safe work practices that includes references to where related information can be found in this book.

GENERAL INDUSTRY CHECKLIST

Yes	No	Inspection Activity	OSHA 29 CFR Reference	NFPA 70E Reference 2000 Ed.	NFPA 70E Reference 2004 Ed.	Program Book Reference	NEC Article Reference
		Installation: Substations, Switchboards, Panels, and Major Equipment					
☐	☐	Is the location (working environment) adequate?	1910.303 (a) & (b)	Part I, 1-3.1 (1)	400.2		110.3
☐	☐	Do labels indicate third-party compliance?	1910.303(b)(i)	Part I, 1-2, 1-3, 1-3.1	400.2, 400.3(A)		110.3(A)
☐	☐	Is equipment installed according to manufacturer's recommendations?	1910.303(b)(2)	Part I, 1-3.2	400.3(B)		110.3(B)
☐	☐	Are all live parts covered?	1910.303(d) & 1910.303(g)(2)	Part I, 1-3.7.1	400.8(A)		110.12(A) & 408.18
☐	☐	Is equipment adequately protected with a barrier or fence?	1910.303(h)(2)	Part I, 1-9.2	400.18		110.26(D) & 110.33(A)(2)
☐	☐	Are exposed live parts accessible to qualified personnel only?	1910.303(h)(2)	Part I, 1-9.1 & 1-9.2.1.1	400.17, 400.17(B)		110.26(D) & 110.33(A)(2) & 110.34(C)
☐	☐	Is working space around equipment adequate?	1910.303(g)(1)(i), Table S-1	Part I, 1-8.1.1.1	400.21		110.26(A) & 110.34(A)
☐	☐	Are dedicated space requirements met?	1910.303(g)(2)(ii)	Part I, 1-8.1.6	400.15(F)		110.26(F)
☐	☐	Do you ensure that no equipment is stored in front of panels?	1910.303(g)(1)(ii)	Part I, 1-8.1.2	400.15(B)		110.26(B)
☐	☐	Is equipment adequately supported?		Part I, 1-3.8	400.9		110.13
☐	☐	Is adequate illumination available?	1910.303(g)(1)(v)	Part I, 1-8.1.4	400.15(D)		110.26(D) & 110.34(D)
☐	☐	Are panel(s) and branch circuits identified?	1910.303(f)	Part I, 1-7	400.14		408.4
☐	☐	Are unused openings covered?	1910.305(b)	Part I, 1-3.7.1	400.8(A)		110.12(A)
☐	☐	Are conductors protected from damage?	1910.304(e)(1)(i)	Part I, 3-1.2.3.8	420.1(B)(2)(g)		300.4
☐	☐	Are cables and conductors terminated correctly?	1910.305(g)(2)(iii)(B) & 1910.305(a)(1)	Part I, 3-1.2.3.2	420.1(B)(2)(b)		110.14(A)
		Installation: Single- and Multiple-Conductor Cables					
☐	☐	Approved for purpose and environment?	1910.303(a) & 1910.303(b)	Part I, 1-2 & 1-3	400.2		110.3(A) & 310.13
☐	☐	Subject to physical damage?		Part I, 3-1.2.3.2	420.1(B)(2)(g)		300.4
☐	☐	Marked as required?	1910.303(b)	Part I, 1-2 & 1-3	400.2, 400.3		310.11
☐	☐	Identified as required?	1910.304(a)	Part I, 3-6	400.7		310.12(B)
☐	☐	Is overcurrent protection adequate for the ampacity?	1910.304(e)(1)(i)	Part I, 2-5.1.1	410.9(A)(1)		240.5(A)

GENERAL INDUSTRY CHECKLIST *(Continued)*

Yes	No	Inspection Activity	OSHA 29 CFR Reference	NFPA 70E Reference 2000 Ed.	NFPA 70E Reference 2004 Ed.	Program Book Reference	NEC Article Reference
		Installation: Flexible Cords and Cables					
☐	☐	Approved for purpose and environment?	1910.303(a) & (b)	Part I, 1-2 & 1-3	400.2, 400.3		110.3(A) & 310.13
☐	☐	Subject to physical damage?		Part I, 3-1.2.3.2	420.1(B)(2) (g)		300.4
☐	☐	Marked as required?	1910.303(b)	Part I, 1-3.1	400.2, 400.3		310.11
☐	☐	Identified as required?	1910.304(a)	Part I, 3-6	400.7		310.12(B)
☐	☐	Is overcurrent protection adequate for the ampacity?	1910.304(e)(1)(i)	Part I, 2-5.1.1	410.9(A)(1)		240.5(A)
☐	☐	Supported as required?		Part I, 3-1.2.3.10	420.1(B)(2) (i)		400.10
☐	☐	Terminated as required?	1910.305(g)(2) (iii)	Part I, 3-1.2.3.9	420.1(B)(2) (h)		400.14
☐	☐	Are cords that pass through holes protected from physical damage?	1910.305(g)(1) (iii)(B)	Part I, 3-1.2.3.8	420.1(B)(2) (g)		400.14
		Installation: Raceways and Cable Trays					
☐	☐	Approved for purpose and environment?	1910.303(a) & (b) & 1910.305(a)(3)	Part I, 1-2, 1-3, 3-1.3.1	400.1, 400.2		110.3(A)
☐	☐	Subject to physical damage?	1910.305(a)(3)(ii)	Part I, 3-1.2.3.2	420.1(B)(2) (g)		Chapter 3
☐	☐	Supported as required?	1910.303(b)(2)	Part I, 3-1.3.1			Chapter 3
		Installation: Junction Boxes					
☐	☐	Suitable for the purpose or environment?	1910.303(a) & (b)	Part I, 1-2 & 1-3	400.2, 400.3		110.3(A) & (B)
☐	☐	Are they secured to the wall or structure?					314.23
☐	☐	Are unused openings covered?	1910.305(b)	Part I, 1-3.7.1	400.8(A)		110.12(A)
☐	☐	Are all live parts covered?	1910.305(b)(2)	Part I, 1-5 & 3-2.1.1	400.12, 420.2(A)(1)		314.25
		Installation: Lighting					
☐	☐	Suitable for the location and environment?	1910.303(a) & (b)	Part I, 1-3	400.2, 400.3		110.3(A) & (B)
☐	☐	Do labels indicate third-party compliance?	1910.303(b)(i)	Part I, 1-2 & 1-6	400.3		110.3(A)
☐	☐	Is equipment installed according to manufacturer's recommendations?	1910.303(b)(2)	Part I, 1-3.2	400.3(B)		110.3(B)
		Installation: Devices					
☐	☐	Are all 15- and 20-amp receptacles protected by GFCIs?		Part I, 2-4.2 & 2-4.2.1	410.4		210.8
☐	☐	Are all receptacles of the grounding type?	1910.305(a)(2) (iii)(C)	Part I, 2-2.2	410.2(B)		406.3(A)

GENERAL INDUSTRY CHECKLIST *(Continued)*

Yes	No	Inspection Activity	OSHA 29 CFR Reference	NFPA 70E Reference 2000 Ed.	NFPA 70E Reference 2004 Ed.	Program Book Reference	NEC Article Reference
		Installation: Devices *(continued)*					
☐	☐	Are receptacles of the required configuration?	1910.305(j)(2)	Part I, 2-2.2.6 & 3-10.2.2	410.2(B)(6)		406.7
☐	☐	Are receptacles in damp locations GFCI protected?	1910.305(j)(2)(ii)	Part I, 3-10.2.3.1	420.10(B)(3)		406.8
		Installation: Generators					
☐	☐	Suitable for the location and environment?	1910.303(b)	Part I, 1-3	400.2, 400.3		110.3(A)
☐	☐	Do labels indicate third-party compliance?	1910.303(b)(i)	Part I, 1-3.2	400.3(B)		110.3(A)
☐	☐	Is overcurrent protection adequate for the equipment and loads?	1910.304(e)(1)(i)	Part I, 2-5.1.1	410.9(A)(1)		445.12
☐	☐	Is a disconnecting means available between the generator and its loads?	1910.304(d)	Part I, 2-4.1.1	410.8(A)(1)		445.18
		Installation: Grounding					
☐	☐	Is the service adequately grounded?	1910.304(f)	Part I, 2-6.1	410.10(A)		250.20
☐	☐	Is any service that is supplied from a separately derived source, such as a transformer or generator, properly grounded?	1910.304(f)(1)(iv) & 1910.304(f)(1)(v)(C)	Part I, 2-6	410.10(C)		250.20
☐	☐	Is the grounding path permanent and continuous?	1910.304(f)(4)	Part I, 2-6.2 & 2-6.3	410.10(C)		250.4(A)(5)
☐	☐	Is the grounding electrode conductor connected to both the equipment grounding conductor and the grounded circuit conductor?	1910.304(f)(3)	Part I, 2-6.2.1	410.10(C)		250.130(A)
☐	☐	Is the grounding conductor adequately sized?	1910.304(f)(3)	Part I, 2-6.3	410.10(C)		250.66
☐	☐	Is the equipment effectively grounded?	1910.304(f)(6)(ii)	Part I, 2-6.5	410.10(E)		250.130
		Installation: Hazardous Locations					
☐	☐	Is the equipment approved for hazardous locations?	1910.303(b) & 1910.307(b)	Part I, 5-2.1 & 5-2.4	440.2(A), 440.2(E)		500.8
☐	☐	Is the equipment marked for the specific occupancy?	1910.307(b)(2)(i)	Part I, 5-2.4	440.2(E)		500.8(B)
☐	☐	Has the classification of the location been determined?	1910.307(a)	Part I, 5-1	440.1		500.5
☐	☐	Are the appropriate wiring methods used?	1910.307(c)	Part I, 5-2.3	440.2(D)		Chapter 5

GENERAL INDUSTRY CHECKLIST *(Continued)*

Yes	No	Inspection Activity	OSHA 29 CFR Reference	NFPA 70E Reference 2000 Ed.	NFPA 70E Reference 2004 Ed.	Program Book Reference	NEC Article Reference
		Safe Work Practices: Electrical Safety Program (ESP)					
☐	☐	Do you have a written ESP?	1910.333	Part II, 1-3	110.3	3, 12	
☐	☐	Who is in charge of the ESP?		Part II, 1-3	110.3	3, 12	
☐	☐	Does the ESP have a budget?				11, App K	
☐	☐	Who maintains the safety program book?				12	
☐	☐	Are qualified people doing the work?	1910.332(b)(3)(iii)	Part II, 2-1.1.2	110.8(B)(3)	10	
☐	☐	Are safety meetings held with the workers?				12	
☐	☐	Who is responsible for the job safety meetings?				12	
☐	☐	Are records kept of the safety meetings (topics, attendees, handouts, etc.)?				12, 13	
☐	☐	Are all procedures audited?				12, 13	
☐	☐	Are the contractors and vendors integrated into the ESP?				12	
☐	☐	Does the company have an electrical safety awareness program?				12	
		Safe Work Practices: Training					
☐	☐	Who is responsible for safety training at the company?				10	
☐	☐	Are the hazards identified and training provided to address the hazards?	1910.332(a)	Part II, 1-5.1	110.6(A)	10	
☐	☐	Are training level requirements for contractors/vendors specified?				13	
☐	☐	Do contractors/vendors provide documentation that their workers are adequately trained?				12, 13	
☐	☐	Are workers trained for work in confined spaces?	1910.332(a)	Part II, 1-5.4	110.6(D)	10	
☐	☐	Is in-house training conducted? How often? Who does the training?				10	
☐	☐	Who maintains the records for the safety training? Where are they kept?				13	

GENERAL INDUSTRY CHECKLIST *(Continued)*

Yes	No	Inspection Activity	OSHA 29 CFR Reference	NFPA 70E Reference 2000 Ed.	NFPA 70E Reference 2004 Ed.	Program Book Reference	NEC Article Reference
Safe Work Practices: Lockout/Tagout (LO/TO)							
☐	☐	Do you have a written LO/TO procedure?	1910.333(b)(2)	Part II, 5-1 & 5-2	120.2	9	
☐	☐	Is the procedure readily available to all workers?				12	
☐	☐	Does the LO/TO procedure cover all tasks?				9	
☐	☐	Does the LO/TO procedure cover all types of energy available in the company?	1910.333(b)(2)(ii)(B)	Part II, 5-1.1	120.2(B)(4)	9	
☐	☐	Are all LO/TO procedures audited at least once every year?	1910.333(b)(2)(iii)	Part II, 5-1.1	120.2(C)(3)	13	
☐	☐	Where are the audit records kept?				13	
☐	☐	Are all required workers trained in LO/TO procedures?				9, 10	
☐	☐	Is LO/TO equipment kept on site?				12	

Safe Work Practices: Written Procedures

Yes	No	Do written procedures exist for the following?	OSHA 29 CFR Reference	2000 Ed.	2004 Ed.	Program Book Reference	NEC Article Reference
☐	☐	The electrical safety program				1, 3, 5, App A	
☐	☐	Electrical drawings updated; Used				5	
☐	☐	Creating an electrically safe work condition				5	
☐	☐	De-energizing electrical equipment				5	
☐	☐	Developing work plans				5, App D	
☐	☐	Lockout/tagout	1910.333(b)(2)			5, 9, App F	
☐	☐	Personal protective grounds (safety grounds)				5, App D	
☐	☐	Selecting and using voltmeters				5	
☐	☐	Purchase and care of personal protective equipment (PPE)				5, 8	
☐	☐	Labeling, marking, and identification				5	
☐	☐	Switching electrical circuits (normal and emergency)				5, App C, D, E	
☐	☐	Electrical test equipment and special tools				5	
☐	☐	Electrically hazardous work permits			130.1(A)	5, App E	
☐	☐	Conducting a hazard/risk analysis			110.8(B)(1)	5, 6, 7	

GENERAL INDUSTRY CHECKLIST *(Continued)*

Yes	No	Do written procedures exist for the following?	OSHA 29 CFR Reference	NFPA 70E Reference 2000 Ed.	NFPA 70E Reference 2004 Ed.	Program Book Reference	NEC Article Reference
		Safe Work Practices: Written Procedures *(continued)*					
☐	☐	Conducting a flash-hazard analysis			130.3	5, 7	
☐	☐	Work on or near energized electrical equipment				5, 6, 7	
☐	☐	Inserting and removing units from energized motor control centers and similar equipment				5, 6, 7, App C	
☐	☐	Work on energized medium-voltage motor control centers				5, 6, 7, App C	
☐	☐	Ground-fault circuit interrupters					
☐	☐	Temporary wiring					527.1
☐	☐	Electric welding machines and portable generators				5, 6, 7, App C	
☐	☐	Cable tray work				5, 6, 7, App C	
☐	☐	Stationary battery installations					
☐	☐	Dismantling and rearranging				5, 6, 7, App C	
☐	☐	Testing and inspecting electrical equipment and cables				5, 6, 7, App C	
☐	☐	Work on 120/240-volt energized equipment				5, 6, 7, App C	
☐	☐	Work on ungrounded electrical circuits				5, 6, 7, App C	
☐	☐	Work on large-capacity DC equipment				5, 6, 7, App C	
☐	☐	Work on variable frequency equipment				5, 6, 7, App C	
☐	☐	Portable electrical equipment				5, 6, 7, App C	
☐	☐	Using mobile equipment near overhead conductors				5, 6, 7, App C	
☐	☐	Safe approach distance				5, 6, 7, App C	
☐	☐	Training and qualification				1, 5,10	
☐	☐	Electrical qualification of contractors				1, 5,10	
☐	☐	Underground cables				5, 6, 7, App C	
☐	☐	Electrical checkout and startup				5, 6, 7, App C	
☐	☐	Electrical accident response				5, 13	
☐	☐	Housekeeping, cleaning, and storage				5, 13	
☐	☐	Auditing and recordkeeping				5, 11, 13	

GENERAL INDUSTRY CHECKLIST *(Continued)*

TASK ASSESSMENTS ON OR NEAR LIVE PARTS

Safe Work Practices: 600-volt switchgear (with power circuit breakers or fused switches)

Yes	No	*Do you perform the following task assessments?*	OSHA 29 CFR Reference	NFPA 70E Reference 2000 Ed.	NFPA 70E Reference 2004 Ed.	Program Book Reference	NEC Article Reference
☐	☐	Operating a circuit breaker or fused switch with the enclosure door closed	1910.333(A) & 1910.335(A)		130.1(A), 130.7	App C, Task A1	
☐	☐	Reading a panel meter while operating a meter switch	1910.333(A) & 1910.335(A)		130.1(A), 130.7	App C, Task A2	
☐	☐	Operating a circuit breaker, fused switch, or starter with the enclosure door open	1910.333(A) & 1910.335(A)		130.1(A), 130.7	App C, Task A3	

Yes	No	*Do written procedures exist for the following?*					
☐	☐	Working on energized parts, including voltage testing	1910.333(A) & 1910.335(A)		130.1(A), 130.7	App C, Task A4	
☐	☐	Working on control circuits with exposed parts energized at 120 volts or below	1910.333(A) & 1910.335(A)		130.1(A), 130.7	App C, Task A5	
☐	☐	Working on control circuits with exposed parts energized at more than 120 volts	1910.333(A) & 1910.335(A)		130.1(A), 130.7	App C, Task A6	
☐	☐	Inserting or removing (racking) circuit breakers from cubicles with the doors open	1910.333(A) & 1910.335(A)		130.1(A), 130.7	App C, Task A7	
☐	☐	Inserting or removing (racking) circuit breakers from cubicles with the doors closed	1910.333(A) & 1910.335(A)		130.1(A), 130.7	App C, Task A8	
☐	☐	Opening hinged covers (to expose bare, energized parts)	1910.333(A) & 1910.335(A)		130.1(A), 130.7	App C, Task A9	
☐	☐	Applying safety grounds after voltage testing	1910.333(A) & 1910.335(A)		130.1(A), 130.7	App C, Task A10	

Safe Work Practices: 600-volt motor control centers

Yes	No	*Do you perform the following task assessments?*					
☐	☐	Operating a circuit breaker or fused switch with the enclosure door closed	1910.333(A) & 1910.335(A)		130.1(A), 130.7	App C, Task B1	
☐	☐	Reading a panel meter while operating a meter switch	1910.333(A) & 1910.335(A)		130.1(A), 130.7	App C, Task B2	

Yes	No	*Do written procedures exist for the following?*					
☐	☐	Operating a circuit breaker, fused switch, or starter with the enclosure door open	1910.333(A) & 1910.335(A)		130.1(A), 130.7	App C, Task B3	

GENERAL INDUSTRY CHECKLIST *(Continued)*

Yes	No	*Do written procedures exist for the following?*	OSHA 29 CFR Reference	NFPA 70E Reference 2000 Ed.	NFPA 70E Reference 2004 Ed.	Program Book Reference	NEC Article Reference

TASK ASSESSMENTS ON OR NEAR LIVE PARTS (continued)

Safe Work Practices: 600-volt motor control centers (continued)

Yes	No	Task	OSHA 29 CFR Reference	2000 Ed.	2004 Ed.	Program Book Reference	NEC
☐	☐	Inserting or removing individual units (buckets) from a motor control center	1910.333(A) & 1910.335(A)		130.1(A), 130.7	App C, Task B4	
☐	☐	Working on energized parts, including voltage testing	1910.333(A) & 1910.335(A)		130.1(A), 130.7	App C, Task B5	
☐	☐	Working on control circuits with exposed parts energized at 120 volts or below	1910.333(A) & 1910.335(A)		130.1(A), 130.7	App C, Task B6	
☐	☐	Working on control circuits with exposed parts energized at more than 120 volts	1910.333(A) & 1910.335(A)		130.1(A), 130.7	App C, Task B7	
☐	☐	Removing bolted covers (to expose bare, energized parts)	1910.333(A) & 1910.335(A)		130.1(A), 130.7	App C, Task B8	
☐	☐	Opening hinged covers (to expose bare, energized parts)	1910.333(A) & 1910.335(A)		130.1(A), 130.7	App C, Task B9	
☐	☐	Applying safety grounds after voltage testing	1910.333(A) & 1910.335(A)		130.1(A), 130.7	App C, Task B10	

Safe Work Practices: Panelboards rated more than 240 volts up to 600 volts

Yes	No	*Do you perform the following task assessments?*	OSHA 29 CFR Reference	2000 Ed.	2004 Ed.	Program Book Reference	NEC
☐	☐	Operating a circuit breaker or fused switch with the cover on	1910.333(A) & 1910.335(A)		130.1(A), 130.7	App C, Task C1	
☐	☐	Operating a circuit breaker or fused switch with the cover off	1910.333(A) & 1910.335(A)		130.1(A), 130.7	App C, Task C2	
☐	☐	Working on energized parts, including voltage testing	1910.333(A) & 1910.335(A)		130.1(A), 130.7	App C, Task C3	

Safe Work Practices: Panelboards rated 240 volts and below

Yes	No	Task	OSHA 29 CFR Reference	2000 Ed.	2004 Ed.	Program Book Reference	NEC
☐	☐	Operating a circuit breaker or fused switch with the cover on	1910.333(A) & 1910.335(A)		130.1(A), 130.7	App C, Task D1	
☐	☐	Operating a circuit breaker or fused switch with the cover off	1910.333(A) & 1910.335(A)		130.1(A), 130.7	App C, Task D2	
☐	☐	Working on energized parts, including voltage testing	1910.333(A) & 1910.335(A)		130.1(A), 130.7	App C, Task D3	
☐	☐	Removing/installing circuit breakers or fused switches	1910.333(A) & 1910.335(A)		130.1(A), 130.7	App C, Task D4	
☐	☐	Removing bolted covers (to expose bare, energized parts)	1910.333(A) & 1910.335(A)		130.1(A), 130.7	App C, Task D5	
☐	☐	Opening hinged covers (to expose bare, energized parts)	1910.333(A) & 1910.335(A)		130.1(A), 130.7	App C, Task D6	

GENERAL INDUSTRY CHECKLIST *(Continued)*

Yes	No	Do you perform the following task assessments?	OSHA 29 CFR Reference	NFPA 70E Reference 2000 Ed.	NFPA 70E Reference 2004 Ed.	Program Book Reference	NEC Article Reference
		TASK ASSESSMENTS ON OR NEAR LIVE PARTS *(continued)*					
		Safe Work Practices: Common tasks on systems rated 600 volts and below					
☐	☐	Removing/replacing light fixture ballast	1910.333(A) & 1910.335(A)		130.1(A), 130.7	App C, Task E1	
☐	☐	Replacing 15- or 20-amp receptacle or switch	1910.333(A) & 1910.335(A)		130.1(A), 130.7	App C, Task E2	
☐	☐	Disconnecting and reconnecting utilization equipment rated at 240 volts or below	1910.333(A) & 1910.335(A)		130.1(A), 130.7	App C, Task E3	
☐	☐	Disconnecting and reconnecting utilization equipment rated at more than 240 volts	1910.333(A) & 1910.335(A)		130.1(A), 130.7	App C, Task E4	
☐	☐	Voltage testing at utilization equipment rated at 240 volts or below	1910.333(A) & 1910.335(A)		130.1(A), 130.7	App C, Task E5	
☐	☐	Voltage testing at utilization equipment rated at more than 240 volts up to 600 volts	1910.333(A) & 1910.335(A)		130.1(A), 130.7	App C, Task E6	
		Safe Work Practices: Other equipment rated 600 volts and below					
☐	☐	Removing bolted covers (to expose bare, energized parts)	1910.333(A) & 1910.335(A)		130.1(A), 130.7	App C, Task F1	
☐	☐	Opening hinged covers (to expose bare, energized parts)	1910.333(A) & 1910.335(A)		130.1(A), 130.7	App C, Task F2	
☐	☐	Inserting or removing equipment (such as a revenue meter)	1910.333(A) & 1910.335(A)		130.1(A), 130.7	App C, Task F3	
☐	☐	Working on energized parts, including voltage testing	1910.333(A) & 1910.335(A)		130.1(A), 130.7	App C, Task F4	
☐	☐	Inserting or removing cable trough or tray covers	1910.333(A) & 1910.335(A)		130.1(A), 130.7	App C, Task F5	
☐	☐	Inserting or removing miscellaneous equipment covers	1910.333(A) & 1910.335(A)		130.1(A), 130.7	App C, Task F6	
☐	☐	Applying safety grounds after testing the voltage	1910.333(A) & 1910.335(A)		130.1(A), 130.7	App C, Task F7	
		Safe Work Practices: Common tasks on metal-clad switchgear rated 1 kV and higher					
☐	☐	Operating a circuit breaker or fused switch with the cover on	1910.333(A) & 1910.335(A)		130.1(A), 130.7	App C, Task G1	
☐	☐	Reading a panel meter while operating a meter switch	1910.333(A) & 1910.335(A)		130.1(A), 130.7	App C, Task G2	
☐	☐	Operating a circuit breaker or fused switch with the cover off	1910.333(A) & 1910.335(A)		130.1(A), 130.7	App C, Task G3	
☐	☐	Working on energized parts, including voltage testing	1910.333(A) & 1910.335(A)		130.1(A), 130.7	App C, Task G4	
☐	☐	Working on control circuits with exposed parts energized at 120 volts or below	1910.333(A) & 1910.335(A)		130.1(A), 130.7	App C, Task G5	

GENERAL INDUSTRY CHECKLIST *(Continued)*

Yes	No	Do you perform the following task assessments?	OSHA 29 CFR Reference	NFPA 70E Reference 2000 Ed.	NFPA 70E Reference 2004 Ed.	Program Book Reference	NEC Article Reference
		TASK ASSESSMENTS ON OR NEAR LIVE PARTS *(continued)*					
		Safe Work Practices: Common tasks on metal-clad switchgear rated at 1 kV and higher					
☐	☐	Working on control circuits with exposed energized parts at more than 120 volts	1910.333(A) & 1910.335(A)		130.1(A), 130.7	App C, Task G6	
☐	☐	Inserting or removing (racking) circuit breakers from cubicles with the doors closed	1910.333(A) & 1910.335(A)		130.1(A), 130.7	App C, Task G7	
☐	☐	Inserting or removing (racking) circuit breakers from cubicles with the doors open	1910.333(A) & 1910.335(A)		130.1(A), 130.7	App C, Task G8	
☐	☐	Removing bolted covers (to expose bare, energized parts)	1910.333(A) & 1910.335(A)		130.1(A), 130.7	App C, Task G9	
☐	☐	Opening hinged covers (to expose bare, energized parts)	1910.333(A) & 1910.335(A)		130.1(A), 130.7	App C, Task G10	
☐	☐	Opening voltage transformer or control power transformer compartments	1910.333(A) & 1910.335(A)		130.1(A), 130.7	App C, Task G11	
☐	☐	Applying safety grounds after voltage testing	1910.333(A) & 1910.335(A)		130.1(A), 130.7	App C, Task G12	
		Safe Work Practices: NEMA E2 motor starter rated from 2.3 kV to 7.2 kV					
☐	☐	Reading a panel meter while operating a meter switch	1910.333(A) & 1910.335(A)		130.1(A), 130.7	App C, Task H1	
☐	☐	Operating a contactor with the enclosure door closed	1910.333(A) & 1910.335(A)		130.1(A), 130.7	App C, Task H2	
☐	☐	Operating a contactor with the enclosure door open	1910.333(A) & 1910.335(A)		130.1(A), 130.7	App C, Task H3	
☐	☐	Working on energized parts, including voltage testing	1910.333(A) & 1910.335(A)		130.1(A), 130.7	App C, Task H4	
☐	☐	Working on control circuits with exposed parts energized at 120 volts or below	1910.333(A) & 1910.335(A)		130.1(A), 130.7	App C, Task H5	
☐	☐	Working on control circuits with exposed parts energized at more than 120 volts	1910.333(A) & 1910.335(A)		130.1(A), 130.7	App C, Task H6	
☐	☐	Inserting or removing (racking) circuit breakers from cubicles with the doors closed	1910.333(A) & 1910.335(A)		130.1(A), 130.7	App C, Task H7	
☐	☐	Inserting or removing (racking) circuit breakers from cubicles with the doors open	1910.333(A) & 1910.335(A)		130.1(A), 130.7	App C, Task H8	
☐	☐	Removing bolted covers (to expose bare, energized parts)	1910.333(A) & 1910.335(A)		130.1(A), 130.7	App C, Task H9	

GENERAL INDUSTRY CHECKLIST *(Continued)*

Yes	No	*Do you perform the following task assessments?*	*OSHA 29 CFR Reference*	*NFPA 70E Reference* 2000 Ed.	*NFPA 70E Reference* 2004 Ed.	*Program Book Reference*	*NEC Article Reference*
		TASK ASSESSMENTS ON OR NEAR LIVE PARTS *(continued)*					
		Safe Work Practices: NEMA E2 motor starter rated from 2.3 kV to 7.2 kV *(continued)*					
☐	☐	Opening hinged covers (to expose bare, energized parts)	1910.333(A) & 1910.335(A)		130.1(A), 130.7	App C, Task H10	
☐	☐	Applying safety grounds after voltage testing	1910.333(A) & 1910.335(A)		130.1(A), 130.7	App C, Task H11	
		Safe Work Practices: Other equipment rated at 1 kV or more					
☐	☐	Operating a switch with the enclosure door closed	1910.333(A) & 1910.335(A)		130.1(A), 130.7	App C, Task I1	
☐	☐	Working on energized parts, including voltage testing	1910.333(A) & 1910.335(A)		130.1(A), 130.7	App C, Task I2	
☐	☐	Removing bolted covers to expose bare, energized parts	1910.333(A) & 1910.335(A)		130.1(A), 130.7	App C, Task I3	
☐	☐	Opening hinged covers to expose bare, energized parts	1910.333(A) & 1910.335(A)		130.1(A), 130.7	App C, Task I4	
☐	☐	Operating an outdoor disconnect switch with a hookstick	1910.333(A) & 1910.335(A)		130.1(A), 130.7	App C, Task I5	
☐	☐	Operating an outdoor disconnect switch (gang operated, from grade)	1910.333(A) & 1910.335(A)		130.1(A), 130.7	App C, Task I6	
☐	☐	Examining insulated cable in a manhole or other confined space	1910.333(A) & 1910.335(A)		130.1(A), 130.7	App C, Task I7	
☐	☐	Examining insulated cable in open air	1910.333(A) & 1910.335(A)		130.1(A), 130.7	App C, Task I8	

For any other task not shown on this table, refer to Chapters 5, 6, 7, and Appendix C in this book.

CONSTRUCTION CHECKLIST

Yes	No	Inspection Activity	OSHA 29 CFR Reference	NFPA 70E Reference 2000 Ed.	NFPA 70E Reference 2004 Ed.	Program Book Reference	NEC Article Reference
		Installation: Substations, Switchboards, Panels, and Major Equipment					
☐	☐	Is the location (working environment) adequate?	1926.403(b)(1)(i)	Part I, 1-3.1 (1)	400.2		110.3
☐	☐	Do labels indicate third-party compliance?	1926.403(b)(2)	Part I, 1-2, 1-3,1-3.1	400.2, 400.3(A)		110.3(A)
☐	☐	Is equipment installed according to manufacturer's recommendations?	1926.403(b)(2)	Part I, 1-3.2	400.3(B)		110.3(B)
☐	☐	Are all live parts covered?	1926.403(f)	Part I, 1-3.7.1	400.8(A)		110.12(A) & 408.18
☐	☐	Is equipment adequately protected with a barrier or fence?	1910.303(h)(2)	Part I, 1-9.2	400.98		110.26(D) & 110.33(A)(2)
☐	☐	Are exposed live parts accessible to qualified personnel only?	1910.303(h)(2)	Part I, 1-9.1 & 1-9.2.1.1	400.17, 400.17(B)		110.26(D) & 110.33(A)(2) & 110.34(C)
☐	☐	Is working space around equipment adequate?	1926.403(i)(1)(i), Table K-1	Part I, 1-8.1.1.1	400.21		110.26(A) & 110.34(A)
☐	☐	Are dedicated space requirements met?		Part I, 1-8.1.6	400.15(F)		110.26(F)
☐	☐	Do you ensure that no equipment is stored in front of panels?	1926.403(i)(1)(ii)	Part I, 1-8.1.2	400.15(B)		110.26(B)
☐	☐	Is equipment adequately supported?	1926.403(d)	Part I, 1-3.8	400.9		110.13
☐	☐	Is adequate illumination available?	1910.303(g)(1)(v)	Part I, 1-8.1.4	400.15(D)		110.26(D) & 110.34(D)
☐	☐	Are panel(s) and branch circuits identified?	1926.403 (h)	Part I, 1-7	400.14		408.4
☐	☐	Are unused openings covered?	1926.405(b)	Part I, 1-3.7.1	400.8(A)		110.12(A)
☐	☐	Are cables and conductors terminated correctly?	1926.405(g)(2)(iv) & (v)	Part I, 3-1.2.3.2	420.1(B) (2)(b)		110.14(A)
		Installation: Single- and Multiple-Conductor Cables, Flexible Cords, and Open Wiring					
☐	☐	Is the installation approved for purpose and environment?	1926.403(b) & 1926.401(b)(1)(i)	Part I, 1-2 & 1-3	400.2		110.3(A)
☐	☐	Is the installation subject to physical damage?	1926.405(a)(2)(ii)	Part I, 3-1.2.3.2	420.1(B) (2)(g)		527.4(A), (B), & (C)
☐	☐	Is the installation marked as required?	1926.403(b)	Part I, 1-2 & 1-3	400.2, 400.3		310.11
☐	☐	Is the installation identified as required?	1926.404(a)	Part I, 3-6	400.7		310.12(B)
☐	☐	Is overcurrent protection adequate for the ampacity?	1926.404(e)	Part I, 2-5.1.1	410.9(A)(1)		240.5(A)
☐	☐	Is the installation supported as required?	1926.405(a)(2)(ii)	Part I, 3-1.2.3.10	420.1(B) (2)(i)		527.4(J)
☐	☐	Is the installation terminated as required?	1926.405(g)(2)(iv)	Part I, 3-1.2.3.9	420.1(B) (2)(h)		110.14(A)

CONSTRUCTION CHECKLIST *(Continued)*

Yes	No	Inspection Activity	OSHA 29 CFR Reference	NFPA 70E Reference 2000 Ed.	NFPA 70E Reference 2004 Ed.	Program Book Reference	NEC Article Reference
		Installation: Single- and Multiple-Conductor Cables, Flexible Cords, and Open Wiring					
☐	☐	Are cords that pass through holes protected from physical damage?	1926.405(g)(2)(v)	Part I, 3-1.2.3.8	420.1(B)(2)(h)		527.4(H)
		Installation: Raceways and Cable Trays					
☐	☐	Is the installation approved for purpose and environment?	1926.403(b) & 1926.405(f)	Part I, 1-2, 1-3, 3-1.3.1	400.1, 400.2		110.3(A)
		Installation: Junction Boxes					
☐	☐	Are the junction boxes suitable for the purpose or environment?	1926.403(b) & 1926.405(e)	Part I, 1-2 & 1-3	400.2, 400.3		110.3(A) & (B)
☐	☐	Are unused openings covered?	1926.405(b)	Part I, 1-3.7.1	400.8(A)		110.12(A)
☐	☐	Are all live parts covered?	1926.403(f) & 1926.405(b)(2)	Part I, 1-5 & 3-2.1.1	400.12, 420.2(A)(1)		314.25
		Installation: Lighting					
☐	☐	Is the lighting suitable for the location and environment?	1926.403(b)	Part I, 1-3	400.2, 400.3		110.3(A) & (B)
☐	☐	Do labels indicate third-party compliance?	1926.403(g)	Part I, 1-2 & 1-6	400.3		110.3(A)
☐	☐	Is equipment installed according to manufacturer's recommendations?	1926.403(b)(2)	Part I, 1-3.2	400.3(B)		110.3(B)
☐	☐	Are all lamps protected from physical damage?	1926.405(a)(2)(ii)(E)	Part I, 1-8.2.2	400.16(B)		527.34(F)
		Installation: Devices					
☐	☐	Are all 15- and 20-amp receptacles protected by GFCIs?	1926.406(b)(1)(ii)	Part I, 2-4.2 & 2-4.2.1	410.4		527.6(A)
☐	☐	Are all receptacles of the grounding type?	1926.405(a)(2)(ii)(C)	Part I, 2-2.2 & 3-1.2.3.3	410.2(B)		527.4(D)
☐	☐	Are receptacles of the required configuration?	1926.405(j)(2)	Part I, 2-2.2.6 & 3-10.2.2	410.2(B)(6)		406.2(F)
☐	☐	Are receptacles in damp locations GFCI protected?	1926.405(j)(2)(ii)	Part I, 3-10.2.3.1	420.10(B)(3)		110.3(A)
		Installation: Grounding					
☐	☐	Is the temporary service adequately grounded?	1926.404(f)	Part I, 2-6.1	410.10(A)		250.20
☐	☐	Is any service that is supplied from a separately derived source, such as a transformer or generator, properly grounded?	1926.404(f)(5)	Part I, 2-6	410.6(C)		250.20
☐	☐	Is the grounding path permanent and continuous?	1926.404(f)(6)	Part I, 2-6.2 & 2-6.3	410.10(C)		250.4(A)(5)

CONSTRUCTION CHECKLIST *(Continued)*

Yes	No	Inspection Activity	OSHA 29 CFR Reference	NFPA 70E Reference 2000 Ed.	NFPA 70E Reference 2004 Ed.	Program Book Reference	NEC Article Reference
		Installation: Grounding					
☐	☐	Is the grounding electrode conductor connected to both the equipment grounding conductor and the grounded circuit conductor?	1926.404(f)(5)	Part I, 2-6.2.1	410.10(C)		250.130(A)
☐	☐	Is the grounding conductor adequately sized?	1926.404(f)(8)(ii)	Part I, 2-6.3	410.10(C)		250.66
☐	☐	Is the equipment effectively grounded?	1926.404(f)(8)(iii)	Part I, 2-6.5	410.10(E)		250.130
		Installation: Bonding					
☐	☐	Are all electrical components bonded together?	1926.404(f)(9)	Part I, 2-6.6 & 3-3.1.1	410.10(B)		250.90
☐	☐	Are the bonding conductors of adequate size?	1926.404(f)(9)	Part I, 2-6.6	410.10(B)		250.122
		Installation: Assured Grounding Program					
☐	☐	Is an assured equipment grounding program in place? (If no, disregard other questions.)	1926.404(b)(1)(iii)	Part I, 2-2.4.2.2.2	410.4(B)(2)(b)		527.6(B)(2)
☐	☐	Is a written copy of the assured equipment grounding program on file?	1926.404(b)(1)(iii)(A)	Part I, 2-2.4.2.2.2	410.4(B)(2)(b)		527.6(B)(2)
☐	☐	Is a designated competent person in charge of the assured equipment grounding program?	1926.404(b)(1)(iii)(B)	Part I, 2-2.4.2.2.2	410.4(B)(2)(b)		527.6(B)(2)
☐	☐	Is each cord set inspected for external defects before use?	1926.404(b)(1)(iii)(C)	Part I, 2-2.4.2.2.2	410.4(B)(2)(b)		527.6(B)(2)
☐	☐	Has each cord set and receptacle been tested before use and at intervals not exceeding 3 months?	1926.404(b)(1)(iii)(D) & (E)	Part I, 2-2.4.2.2.2(3)	410.4(B)(2)(b)		527.6(B)(2)
		Installation: Wiring					
☐	☐	Are feeders adequately supported?	1926.405(a)(2)(ii)	Part I, 3-1.2.3.2	420.1(B)		527.6(B)(2)
☐	☐	Are all extension cords used with portable tools of the 3-wire and "extra-hard use" type?	1926.405(a)(2)(ii)(J)				527.4(C)
		Installation: Hazardous Locations					
☐	☐	Is the equipment approved for hazardous locations?	1910.303(b) & 1910.307(b)	Part I, 5-2.1 & 5-2.4	440.2(A), 440.2(E)		500.8
☐	☐	Is the equipment marked for the specific occupancy?	1910.307(b)(2)(i)	Part I, 5-2.4	440.2(E)		500.8(B)

CONSTRUCTION CHECKLIST *(Continued)*

Yes	No	Inspection Activity	OSHA 29 CFR Reference	NFPA 70E Reference 2000 Ed.	NFPA 70E Reference 2004 Ed.	Program Book Reference	NEC Article Reference
		Installation: Hazardous Locations *(continued)*					
☐	☐	Has the classification of the location been determined?	1910.307(a)	Part I, 5-1	440.1		500.5
☐	☐	Are the appropriate wiring methods used?	1910.307(c)	Part I, 5-2.3	440.2(D)		Chapter 5
		Safe Work Practices: Electrical Safety Program					
☐	☐	Do you have a written electrical safety program?	1926.20(b)	Part II, 2-3	110.3	2, 13	
☐	☐	Is a competent person in charge?	1926.21(b)(2)	Part II, 1-3	110.3	2	
☐	☐	Are qualified people doing the work?	1926.21(b)(4)	Part II, 2-1.1.2	110.8(B)(3)	10	
		Safe Work Practices: Training					
☐	☐	Are the hazards identified and is training provided to address the hazards?	1926.21(b)(2)	Part II, 1-5.1	110.6(A)	5, 6, 7, 10	
☐	☐	Are workers trained for work in confined spaces?	1926.20(b)(6)(i)	Part I, 1-5.4	110.6(A)	10	
☐	☐	Is documentation of the training maintained?	1926.20(b)	Part II, 1-3	110.3	10, 13	
		Safe Work Practices: Lockout/Tagout					
☐	☐	Do you have a lockout/tagout program?	1926.20(b)	Part II, 5-1 & 5-2	120.2	5, 9	
☐	☐	Is all equipment that is locked/tagged out rendered inoperative?	1926.417(b)	Part II, 5-1.1	120.2(B)(4)	9	
☐	☐	Is a tag installed on all locked/tagged out equipment?	1926.417(c)	Part II, 5-1.1	120.2(C)(3)	9	

Safe Work Practices: Written Procedures

Yes	No	Do written procedures exist for the following?	Program Book Reference
☐	☐	The electrical safety program	1, 3, 5, App A
☐	☐	Creating an electrically safe work condition	5
☐	☐	De-energizing electrical equipment	5
☐	☐	Developing work plans	5, App D
☐	☐	Lockout/tagout	5, 9, App F
☐	☐	Personal protective grounds (safety grounds)	5, App D
☐	☐	Selecting and using voltmeters	5

CONSTRUCTION CHECKLIST *(Continued)*

Yes	No	Do written procedures exist for the following?	OSHA 29 CFR Reference	NFPA 70E Reference 2000 Ed.	NFPA 70E Reference 2004 Ed.	Program Book Reference	NEC Article Reference
		Safe Work Practices: Written Procedures *(continued)*					
☐	☐	Purchase and care of personal protective equipment (PPE)				5, 8	
☐	☐	Labeling, marking, and identification				5	
☐	☐	Switching electrical circuits (normal and emergency)				5, App C	
☐	☐	Electrical test equipment and special tools				5	
☐	☐	Electrically hazardous work permits				5, App E	
☐	☐	Conducting a hazard/risk analysis				5, 6, 7	
☐	☐	Conducting a flash-hazard analysis				5, 7	
☐	☐	Work on or near energized electrical equipment				5, 6, 7	
☐	☐	Work on or near energized conductors and circuit parts				5, 6, 7	
☐	☐	Inserting and removing units from energized motor control centers and similar equipment				5, 6, 7, App C	
☐	☐	Work on energized medium-voltage motor control centers				5, 6, 7, App C	
☐	☐	Ground-fault circuit interrupters					
☐	☐	Temporary wiring					
☐	☐	Electric welding machines and portable generators				5, 6, 7, App C	
☐	☐	Cable tray work				5, 6, 7, App C	
☐	☐	Stationary battery installations					
☐	☐	Dismantling and rearranging				5, 6, 7, App C	
☐	☐	Testing and inspecting electrical equipment and cables				5, 6, 7, App C	
☐	☐	Work on 120/240-volt energized equipment				5, 6, 7, App C	
☐	☐	Work on ungrounded electrical circuits				5, 6, 7, App C	
☐	☐	Work on large-capacity DC equipment				5, 6, 7, App C	
☐	☐	Work on variable frequency equipment				5, 6, 7, App C	
☐	☐	Portable electrical equipment				5, 6, 7, App C	
☐	☐	Using mobile equipment near overhead conductors				5, 6, 7, App C	

CONSTRUCTION CHECKLIST (Continued)

Yes	No	Do written procedures exist for the following?	OSHA 29 CFR Reference	NFPA 70E Reference 2000 Ed.	NFPA 70E Reference 2004 Ed.	Program Book Reference	NEC Article Reference
		Safe Work Practices: Written Procedures (continued)					
☐	☐	Safe approach distance				5, 6, 7, App C	
☐	☐	Training and qualification				1, 5,10	
☐	☐	Electrical qualification of contractors				1, 5,10	
☐	☐	Underground cables				5, 6, 7, App C	
☐	☐	Electrical checkout and startup				5, 6, 7, App C	
☐	☐	Electrical accident response				5, 13	
☐	☐	Housekeeping, cleaning, and storage				5, 13	
☐	☐	Working on or near live parts					

Safe Work Practices: Panelboards rated more than 240 volts up to 600 volts

Yes	No	Do you perform the following task assessments?	OSHA 29 CFR Reference	NFPA 70E 2000 Ed.	NFPA 70E 2004 Ed.	Program Book Reference	NEC Article Reference
☐	☐	Operating a circuit breaker or fused switch with the cover on	1926.416(A)		130.1(A), 130.7	App C, Task C1	
☐	☐	Operating a circuit breaker or fused switch with the cover off	1926.416(A)		130.1(A), 130.7	App C, Task C2	
☐	☐	Working on energized parts, including voltage testing	1926.416(A)		130.1(A), 130.7	App C, Task C3	

Safe Work Practices: Panelboards rated 40 volts and below

Yes	No		OSHA 29 CFR Reference	NFPA 70E 2000 Ed.	NFPA 70E 2004 Ed.	Program Book Reference	NEC Article Reference
☐	☐	Operating a circuit breaker or fused switch with the cover on	1926.416(A)		130.1(A), 130.7	App C, Task D1	
☐	☐	Operating a circuit breaker or fused switch with the cover off	1926.416(A)		130.1(A), 130.7	App C, Task D2	
☐	☐	Working on energized parts, including voltage testing	1926.416(A)		130.1(A), 130.7	App C, Task D3	
☐	☐	Removing/installing circuit breakers or fused switches	1926.416(A)		130.1(A), 130.7	App C, Task D4	
☐	☐	Removing bolted covers (to expose bare, energized parts)	1926.416(A)		130.1(A), 130.7	App C, Task D5	

Safe Work Practices: Common tasks on equipment rated 600 volts and below

Yes	No		OSHA 29 CFR Reference	NFPA 70E 2000 Ed.	NFPA 70E 2004 Ed.	Program Book Reference	NEC Article Reference
☐	☐	Removing/replacing light fixture ballast	1926.416(A)		130.1(A), 130.7	App C, Task E1	
☐	☐	Replacing 15- or 20-amp receptacle or switch	1926.416(A)		130.1(A), 130.7	App C, Task E2	
☐	☐	Disconnecting and reconnecting utilization equipment rated at 240 volts or below	1926.416(A)		130.1(A), 130.7	App C, Task E3	

CONSTRUCTION CHECKLIST *(Continued)*

Yes	No	*Do written procedures exist for the following?*	OSHA 29 CFR Reference	NFPA 70E Reference 2000 Ed.	NFPA 70E Reference 2004 Ed.	Program Book Reference	NEC Article Reference
		Safe Work Practices: Common tasks on equipment rated 600 volts and below (continued)					
☐	☐	Disconnecting and reconnecting utilization equipment rated at more than 240 volts	1926.416(A)		130.1(A), 130.7	App C, Task E4	
☐	☐	Voltage testing at utilization equipment rated at 240 volts or below	1926.416(A)		130.1(A), 130.7	App C, Task E5	
☐	☐	Voltage testing at utilization equipment rated at more than 240 volts	1926.416(A)		130.1(A), 130.7	App C, Task E6	
		Safe Work Practices: Other equipment rated 600 volts and below					
☐	☐	Removing bolted covers (to expose bare, energized parts)	1926.416(A)		130.1(A), 130.7	App C, Task F1	
☐	☐	Opening hinged covers (to expose bare, energized parts)	1926.416(A)		130.1(A), 130.7	App C, Task F2	
☐	☐	Inserting or removing equipment (such as a revenue meter)	1926.416(A)		130.1(A), 130.7	App C, Task F3	
☐	☐	Inserting or removing cable trough or tray covers	1926.416(A)		130.1(A), 130.7	App C, Task F5	
☐	☐	Inserting or removing miscellaneous equipment covers	1926.416(A)		130.1(A), 130.7	App C, Task F6	
☐	☐	Applying safety grounds after testing the voltage	1926.416(A)		130.1(A), 130.7	App C, Task F7	

For any other task not shown on this table, refer to Chapters 5, 6, 7, and Appendix C.

　　

Task Assessment Checklists

This appendix contains specific task assessment checklists that can be used in conjunction with the site assessment step (see Chapter 5, Site Assessment) in the development of a company's electrical safety program. These task assessments have been formulated based on Section 130.3(A) and Table 130.2(C) of NFPA 70E, *Standard for Electrical Safety in the Workplace,* 2004 edition, and Appendix G of this book.

Each checklist contains 10 items that are to be evaluated for the specific task and equipment involved. To help clarify the evaluation process and use of the checklists, the following is a detailed description of each item.

Item 1—This item asks the question: Is the equipment operating at more than 50 volts, or is a shock hazard present? Section 130.1 of NFPA 70E, 2004 edition, and 29 CFR 1910.333(a)(1) require that when an employee is exposed to live parts greater than 50 volts, the circuit shall be de-energized before the employee works on or near them, unless the employer can demonstrate that de-energizing introduces additional or increased hazards or is infeasible, due to equipment design or operational limitations. 29 CFR 1910.333(a)(2) requires that if live parts are not deenergized, other safety-related work practices shall be used to protect employees who might be exposed to the electrical hazard involved. If the hazard exists, a task assessment is required.

Item 2—This item determines that a flash hazard analysis is required before a person approaches any exposed electrical conductor or circuit part that has not been placed in an electrically safe work condition.

Item 3—This item determines the shock boundary, which is equal to the restricted approach boundary in Table 130.2(C) of NFPA 70E, 2004 edition. Column 4 of the table provides that information.

Item 4—This item determines that the standard flash protection boundary on systems 600 volts or less is 4 ft, 0 in., based on the information in Section 130.3(A) of NFPA 70E, 2004 edition. Prescriptive criteria are used to reach that conclusion. The user may elect to use alternate calculation methods to determine this boundary.

Item 5—This item selects the hazard/risk category for the task using the applicable task from Appendix G. If the task is not shown in the table, a task assessment should be developed by the company. In 2nd column, a number (0, 1, 2, 2*, 3, or 4) correlates with the columns in Appendix H, which covers personal protective equipment (PPE).

Item 6—The hazard/risk category tables (Appendix G) were determined based on specific incident energy exposure levels. For example, the two principal fault-energy considerations are 25 kA with a 2-cycle fault-clearing time and 65 kA with a 2-cycle clearing time. The two major components for arc-flash exposure are the fault energy available and the amount of time it takes for the overcurrent protective device to clear the fault.

Item 7—This item determines the fault energy level at the location. The available fault energy is defined by this range. If the value exceeds the number on the task assessment checklist, then an arc-flash analysis must be performed.

Item 8—This item identifies additional requirements that may be needed. The legend has three notes that are at the bottom of Appendix G that the user should consider when selecting PPE. The first note requires voltage-rated gloves, rated and tested for the maximum line-to-line voltage upon which work will be done. The second note requires voltage-rated tools, rated and tested for the maximum line-to-line voltage upon which work will be done. The third note specifies that when a hazard/risk category is 2*, a double-layer switching hood and hearing protection are required, in addition to the other hazard/risk Category 2 requirements of Appendix H.

Item 9—This item provides the shock protection for the worker, specifically, the need for voltage-rated tools and voltage-rated gloves. If these items are required, they must be rated and tested for the maximum line-to-line voltage upon which work will be done.

Item 10—This item describes the arc-flash and blast protection for the worker. Under the list of required PPE, notes are provided next to the specific PPE item. These notes provide clarity or alternative methods in consideration of the selection of PPE. See Appendix H for notes.

Within this appendix, the task assessment checklists are grouped by specific equipment types and voltages for easy reference, as follows:

600-Volt Switchgear:

Task A1—Operating a Circuit Breaker or Fused Switch with the Enclosure Door Closed
Task A2—Reading a Panel Meter While Operating a Meter Switch
Task A3—Operating a Circuit Breaker, Fused Switch, or Starter with the Enclosure Door Open
Task A4—Working on Energized Parts, Including Voltage Testing
Task A5—Working on Control Circuits with Exposed Parts Energized at 120 Volts or Below
Task A6—Working on Control Circuits with Exposed Parts Energized at More Than 120 Volts
Task A7—Inserting or Removing (Racking) Circuit Breakers from Cubicles with the Doors Open
Task A8—Inserting or Removing (Racking) Circuit Breakers from Cubicles with the Doors Closed
Task A9—Opening Hinged Covers (to Expose Bare, Energized Parts)
Task A10—Applying Safety Grounds after Voltage Testing

600-Volt Motor Control Centers:

Task B1—Operating a Circuit Breaker or Fused Switch with the Enclosure Door Closed
Task B2—Reading a Panel Meter While Operating a Meter Switch
Task B3—Operating a Circuit Breaker, Fused Switch, or Starter with the Enclosure Door Open
Task B4—Inserting or Removing Individual Units (Buckets) from a Motor Control Center
Task B5—Working on Energized Parts, Including Voltage Testing
Task B6—Working on Control Circuits with Exposed Parts Energized at 120 Volts or Below
Task B7—Working on Control Circuits with Exposed Parts Energized at More Than 120 Volts
Task B8—Removing Bolted Covers (to Expose Bare, Energized Parts)
Task B9—Opening Hinged Covers (to Expose Bare, Energized Parts)
Task B10—Applying Safety Grounds after Voltage Testing

Panelboards Rated More Than 240 Volts up to 600 Volts:

Task C1—Operating a Circuit Breaker or Fused Switch with the Cover On
Task C2—Operating a Circuit Breaker or Fused Switch with the Cover Off
Task C3—Working on Energized Parts, Including Voltage Testing

Panelboards Rated 240 Volts and Below:

Task D1—Operating a Circuit Breaker or Fused Switch with the Cover On
Task D2—Operating a Circuit Breaker or Fused Switch with the Cover Off
Task D3—Working on Energized Parts, Including Voltage Testing
Task D4—Removing/Installing Circuit Breakers or Fused Switches
Task D5—Removing Bolted Covers (to Expose Bare, Energized Parts)
Task D6—Opening Hinged Covers (to Expose Bare, Energized Parts)

Common Tasks on Systems Rated 600 Volts and Below:

Task E1—Removing/Replacing Light Fixture Ballast
Task E2—Replacing 15- or 20-Amp Receptacle or Switch
Task E3—Disconnecting/Reconnecting Utilization Equipment Rated at 240 Volts or Less
Task E4—Disconnecting/Reconnecting Utilization Equipment Rated More Than 240 Volts up to 600 Volts
Task E5—Voltage Testing at Utilization Equipment Rated 240 Volts and Below
Task E6—Voltage Testing at Utilization Equipment Rated More Than 240 Volts up to 600 Volts

Other Equipment Rated at 600 Volts:

Task F1—Removing Bolted Covers (to Expose Bare, Energized Parts)
Task F2—Opening Hinged Covers (to Expose Bare, Energized Parts)

Task F3—Inserting or Removing Equipment (Such as a Revenue Meter)
Task F4—Working on Energized Parts, Including Voltage Testing
Task F5—Inserting or Removing Cable Trough or Tray Covers
Task F6—Installing or Removing Miscellaneous Equipment Covers
Task F7—Applying Safety Grounds after Voltage Testing

Metal-Clad Switchgear Rated at 1 kV and Higher:

Task G1—Operating a Circuit Breaker or Fused Switch with the Enclosure Door Closed
Task G2—Reading a Panel Meter While Operating a Meter Switch
Task G3—Operating a Circuit Breaker, Fused Switch, or Starter with the Enclosure Door Open
Task G4—Working on Energized Parts, Including Voltage Testing
Task G5—Working on Control Circuits with Exposed Parts Energized at 120 Volts or Below
Task G6—Working on Control Circuits with Exposed Parts Energized at More Than 120 Volts
Task G7—Inserting or Removing (Racking) Circuit Breakers from Cubicles with the Doors Closed
Task G8—Inserting or Removing (Racking) Circuit Breakers from Cubicles with the Doors Open
Task G9—Removing Bolted Covers (to Expose Bare, Energized Parts)
Task G10—Opening Hinged Covers (to Expose Bare, Energized Parts)
Task G11—Opening Voltage Transformer or Control Power Transformer Compartments
Task G12—Applying Safety Grounds after Voltage Testing

NEMA E2 Motor Starter Rated from 2.3 kV to 7.2 kV:

Task H1—Reading a Panel Meter While Operating a Meter Switch
Task H2—Operating a Contactor with the Enclosure Door Closed
Task H3—Operating a Contactor with the Enclosure Door Open
Task H4—Working on Energized Parts, Including Voltage Testing
Task H5—Working on Control Circuits with Exposed Parts Energized at 120 Volts or Below
Task H6—Working on Control Circuits with Exposed Parts Energized at More Than 120 Volts
Task H7—Inserting or Removing (Racking) Circuit Breakers from Cubicles with the Doors Open
Task H8—Inserting or Removing (Racking) Circuit Breakers from Cubicles with the Doors Closed
Task H9—Removing Bolted Covers (to Expose Bare, Energized Parts)
Task H10—Opening Hinged Covers (to Expose Bare, Energized Parts)
Task H11—Applying Safety Grounds after Voltage Testing

Other Equipment Rated at 1 kV or More:

Task I1—Operating a Switch with the Enclosure Door Closed
Task I2—Working on Energized Parts, Including Voltage Testing

TASK ASSESSMENT CHECKLIST FOR 600-VOLT CLASS SWITCHGEAR EQUIPMENT
Job Task A1: Operating a Circuit Breaker or Fused Switch with the Enclosure Door Closed

Date: _______________________   **Job Location:** ____

Yes   No

☑   ☐   **1.** Is the equipment operating at more that 50 volts or is a shock hazard present?

If "yes," perform a hazard/risk analysis.

If "no," coordinate with the Electrical Safety Program.

☑   ☐   **2.** Is a flash-hazard analysis required? (70E-2000, 2-1.3.3; 70E-2004, 130.3)

3. Determine the shock protection boundary. (70E-2000 Table 2-1.3.4; 70E-2004 Table 130.2(B))

The restricted approach boundary distance is 1 ft, 0 in.

4. Determine the flash protection boundary. (70E-2000, 2-1.3.3.2; 70E-2004, 130.3(A)(1))

The flash protection boundary distance is 4 ft, 0 in.

5. Select the hazard/risk category for the task from Appendix G.

The hazard risk category is 0.

6. Check notes on the table in Appendix G to determine the incident energy level for the specific task in the table.

☑   ☐   **7.** Is the fault energy level available at the location equal to or less than the table notes in Appendix G?

Note: The available fault current must be 65 kA or less, with 2-cycle clearing time to be applicable.

If "yes," the table is applicable.   If "no," further investigation is required.

8. Check the legend in Appendix G for additional requirements. List requirements:

_______________________________   _______________________________

_______________________________   _______________________________

_______________________________   _______________________________

9. Determine the need for voltage-rated tools and voltage-rated gloves.

☐   ☑   Are voltage-rated tools required?

☐   ☑   Are voltage-rated gloves required?

10. On 70E-2000 Table 3-3.9.2; 70E-2004 Table 130.7(C)(10) locate the hazard/risk category number. Select the PPE based on the table requirements. Read and comply with all applicable notes to ensure safety.

PPE required:

Untreated cotton long-sleeved shirt

Untreated cotton long pants

Safety glasses

                                   (p. 1 of 10)

TASK ASSESSMENT CHECKLIST FOR 600-VOLT CLASS SWITCHGEAR EQUIPMENT
Job Task A2: Reading a Panel Meter While Operating a Meter Switch

Date: _________________________   **Job Location:** _________________________________

Yes   No

☑ ☐   **1.** Is the equipment operating at more that 50 volts or is a shock hazard present?

  If "yes," perform a hazard/risk analysis.

  If "no," coordinate with the Electrical Safety Program.

☑ ☐   **2.** Is a flash-hazard analysis required? (70E-2000, 2-1.3.3; 70E-2004, 130.3)

  3. Determine the shock protection boundary. (70E-2000 Table 2-1.3.4; 70E-2004 Table 130.2(B))

  The restricted approach boundary distance is 1 ft, 0 in.

  4. Determine the flash protection boundary. (70E-2000, 2-1.3.3.2; 70E-2004, 130.3(A)(1))

  The flash protection boundary distance is 4 ft, 0 in.

  5. Select the hazard/risk category for the task from Appendix G.

  The hazard risk category is 0.

  6. Check notes on the table in Appendix G to determine the incident energy level for the specific task in the table.

☑ ☐   **7.** Is the fault energy level available at the location equal to or less than the table notes in Appendix G?

  Note: *The available fault current must be 65 kA or less, with 2-cycle clearing time to be applicable.*

  If "yes," the table is applicable.   If "no," further investigation is required.

  8. Check the legend in Appendix G for additional requirements. List requirements:

  ________________________   ________________________

  ________________________   ________________________

  ________________________   ________________________

  9. Determine the need for voltage-rated tools and voltage-rated gloves.

☐ ☑   Are voltage-rated tools required?

☐ ☑   Are voltage-rated gloves required?

  10. On 70E-2000 Table 3-3.9.2; 70E-2004 Table 130.7(C)(10) locate the hazard/risk category number. Select the PPE based on the table requirements. Read and comply with all applicable notes to ensure safety.

  PPE required:

  Untreated cotton long-sleeved shirt

  Untreated cotton long pants

  Safety glasses

                                                   (p. 2 of 10)

TASK ASSESSMENT CHECKLIST FOR 600-VOLT CLASS SWITCHGEAR EQUIPMENT
Job Task A3: Operating a Circuit Breaker, Fused Switch, or Starter with the Enclosure Door Open

Date: _________________________   Job Location: _______________________________________

Yes   *No*

☑   ☐   **1.** Is the equipment operating at more that 50 volts or is a shock hazard present?

If "yes," perform a hazard/risk analysis.

If "no," coordinate with the Electrical Safety Program.

☑   ☐   **2.** Is a flash-hazard analysis required? (70E-2000, 2-1.3.3; 70E-2004, 130.3)

3. Determine the shock protection boundary. (70E-2000 Table 2-1.3.4; 70E-2004 Table 130.2(B))

The restricted approach boundary distance is 1 ft, 0 in.

4. Determine the flash protection boundary. (70E-2000, 2-1.3.3.2; 70E-2004, 130.3(A)(1))

The flash protection boundary distance is 4 ft, 0 in.

5. Select the hazard/risk category for the task from Appendix G.

The hazard risk category is 1.

6. Check notes on the table in Appendix G to determine the incident energy level for the specific task in the table.

☑   ☐   **7.** Is the fault energy level available at the location equal to or less than the table notes in Appendix G?

Note: The available fault current must be 65 kA or less, with 2-cycle clearing time to be applicable.

If "yes," the table is applicable.   If "no," further investigation is required.

8. Check the legend in Appendix G for additional requirements. List requirements:

________________________________   ________________________________

________________________________   ________________________________

________________________________   ________________________________

9. Determine the need for voltage-rated tools and voltage-rated gloves.

☐   ☑   Are voltage-rated tools required?

☐   ☑   Are voltage-rated gloves required?

10. On 70E-2000 Table 3-3.9.2; 70E-2004 Table 130.7(C)(10) locate the hazard/risk category number. Select the PPE based on the table requirements. Read and comply with all applicable notes to ensure safety.

PPE required:

Untreated cotton long pants (Note 4)	Hard hat
FR long-sleeved shirt	Safety glasses
FR pants (Note 4)	Leather gloves, as needed
Coveralls (Note 5)	Leather shoes, as needed

                                                     (p. 3 of 10)

TASK ASSESSMENT CHECKLIST FOR 600-VOLT CLASS SWITCHGEAR EQUIPMENT
Job Task A4: Working on Energized Parts, Including Voltage Testing

Date: _________________________   **Job Location:** _______________________________

Yes   No

☑   ☐   **1.** Is the equipment operating at more that 50 volts or is a shock hazard present?

If "yes," perform a hazard/risk analysis.

If "no," coordinate with the Electrical Safety Program.

☑   ☐   **2.** Is a flash-hazard analysis required? (70E-2000, 2-1.3.3; 70E-2004, 130.3)

3. Determine the shock protection boundary. (70E-2000 Table 2-1.3.4; 70E-2004 Table 130.2(B))

The restricted approach boundary distance is 1 ft, 0 in.

4. Determine the flash protection boundary. (70E-2000, 2-1.3.3.2; 70E-2004, 130.3(A)(1))

The flash protection boundary distance is 4 ft, 0 in.

5. Select the hazard/risk category for the task from Appendix G.

The hazard risk category is 2.*

6. Check notes on the table in Appendix G to determine the incident energy level for the specific task in the table.

☑   ☐   **7.** Is the fault energy level available at the location equal to or less than the table notes in Appendix G?

Note: The available fault current must be 65 kA or less, with 2-cycle clearing time to be applicable.

If "yes," the table is applicable.   If "no," further investigation is required.

8. Check the legend in Appendix G for additional requirements. List requirements:

_______________________________   _______________________________

_______________________________   _______________________________

_______________________________   _______________________________

9. Determine the need for voltage-rated tools and voltage-rated gloves.

☑   ☐   Are voltage-rated tools required?

☑   ☐   Are voltage-rated gloves required?

10. On 70E-2000 Table 3-3.9.2; 70E-2004 Table 130.7(C)(10) locate the hazard/risk category number. Select the PPE based on the table requirements. Read and comply with all applicable notes to ensure safety.

PPE required:

Untreated cotton t-shirt	Safety glasses or goggles
Untreated cotton long pants (Note 6)	Double layer switching hood
FR long-sleeved shirt (Note 9)	Hearing protection
FR long pants (Note 6)	Leather gloves
Coveralls (Note 7)	Leather shoes
Hard hat	

TASK ASSESSMENT CHECKLIST FOR 600-VOLT CLASS SWITCHGEAR EQUIPMENT
Job Task A5: Working on Control Circuits with Exposed Parts Energized at 120 Volts or Below

Date: ________________________   Job Location: ________________________________

Yes   No

☑   ☐   **1.** Is the equipment operating at more that 50 volts or is a shock hazard present?

If "yes," perform a hazard/risk analysis.

If "no," coordinate with the Electrical Safety Program.

☑   ☐   **2.** Is a flash-hazard analysis required? (70E-2000, 2-1.3.3; 70E-2004, 130.3)

3. Determine the shock protection boundary. (70E-2000 Table 2-1.3.4; 70E-2004 Table 130.2(B))

The restricted approach boundary distance is 1 ft, 0 in.

4. Determine the flash protection boundary. (70E-2000, 2-1.3.3.2; 70E-2004, 130.3(A)(1))

The flash protection boundary distance is 4 ft, 0 in.

5. Select the hazard/risk category for the task from Appendix G.

The hazard risk category is 0.

6. Check notes on the table in Appendix G to determine the incident energy level for the specific task in the table.

☑   ☐   **7.** Is the fault energy level available at the location equal to or less than the table notes in Appendix G?

Note: *The available fault current must be 65 kA or less, with 2-cycle clearing time to be applicable.*

If "yes," the table is applicable.   If "no," further investigation is required.

8. Check the legend in Appendix G for additional requirements. List requirements:

_______________________________   _______________________________

_______________________________   _______________________________

_______________________________   _______________________________

9. Determine the need for voltage-rated tools and voltage-rated gloves.

☑   ☐   Are voltage-rated tools required?

☑   ☐   Are voltage-rated gloves required?

10. On 70E-2000 Table 3-3.9.2; 70E-2004 Table 130.7(C)(10) locate the hazard/risk category number. Select the PPE based on the table requirements. Read and comply with all applicable notes to ensure safety.

PPE required:

Untreated cotton long-sleeved shirt

Untreated cotton long pants

Safety glasses

    (p. 5 of 10)

TASK ASSESSMENT CHECKLIST FOR 600-VOLT CLASS SWITCHGEAR EQUIPMENT

Job Task A6: Working on Control Circuits with Exposed Parts Energized at More Than 120 Volts

Date: ___________________   **Job Location:** _______________________________________

Yes   No

☑ ☐   **1.** Is the equipment operating at more that 50 volts or is a shock hazard present?

      If "yes," perform a hazard/risk analysis.

      If "no," coordinate with the Electrical Safety Program.

☑ ☐   **2.** Is a flash-hazard analysis required? (70E-2000, 2-1.3.3; 70E-2004, 130.3)

      3. Determine the shock protection boundary. (70E-2000 Table 2-1.3.4; 70E-2004 Table 130.2(B))

      The restricted approach boundary distance is 1 ft, 0 in.

      4. Determine the flash protection boundary. (70E-2000, 2-1.3.3.2; 70E-2004, 130.3(A)(1))

      The flash protection boundary distance is 4 ft, 0 in.

      5. Select the hazard/risk category for the task from Appendix G.

      The hazard risk category is 2.*

      6. Check notes on the table in Appendix G to determine the incident energy level for the specific task in the table.

☑ ☐   **7.** Is the fault energy level available at the location equal to or less than the table notes in Appendix G?

      Note: *The available fault current must be 65 kA or less, with 2-cycle clearing time to be applicable.*

      If "yes," the table is applicable.   If "no," further investigation is required.

      8. Check the legend in Appendix G for additional requirements. List requirements:

      ___________________________   ___________________________

      ___________________________   ___________________________

      ___________________________   ___________________________

      9. Determine the need for voltage-rated tools and voltage-rated gloves.

☑ ☐   Are voltage-rated tools required?

☑ ☐   Are voltage-rated gloves required?

      10. On 70E-2000 Table 3-3.9.2; 70E-2004 Table 130.7(C)(10) locate the hazard/risk category number. Select the PPE based on the table requirements. Read and comply with all applicable notes to ensure safety.

      PPE required:

Untreated cotton t-shirt	Safety glasses or goggles
Untreated cotton long pants (Note 6)	Double layer switching hood
FR long-sleeved shirt (Note 9)	Hearing protection
FR long pants (Note 6)	Leather gloves
Coveralls (Note 7)	Leather shoes
Hard hat	

TASK ASSESSMENT CHECKLIST FOR 600-VOLT CLASS SWITCHGEAR EQUIPMENT
Job Task A7: Inserting or Removing (Racking) Circuit Breakers from Cubicles with the Doors Open

Date: ________________________   **Job Location:** ________________________

Yes **No**

☑ ☐ **1.** Is the equipment operating at more that 50 volts or is a shock hazard present?

    If "yes," perform a hazard/risk analysis.

    If "no," coordinate with the Electrical Safety Program.

☑ ☐ **2.** Is a flash-hazard analysis required? (70E-2000, 2-1.3.3; 70E-2004, 130.3)

    3. Determine the shock protection boundary. (70E-2000 Table 2-1.3.4; 70E-2004 Table 130.2(B))
    The restricted approach boundary distance is 1 ft, 0 in.

    4. Determine the flash protection boundary. (70E-2000, 2-1.3.3.2; 70E-2004, 130.3(A)(1))
    The flash protection boundary distance is 4 ft, 0 in.

    5. Select the hazard/risk category for the task from Appendix G.
    The hazard risk category is 3.

    6. Check notes on the table in Appendix G to determine the incident energy level for the specific task in the table.

☑ ☐ **7.** Is the fault energy level available at the location equal to or less than the table notes in Appendix G?
    Note: The available fault current must be 65 kA or less, with 2-cycle clearing time to be applicable.
    If "yes," the table is applicable.   If "no," further investigation is required.

    8. Check the legend in Appendix G for additional requirements. List requirements:

    ________________________   ________________________

    ________________________   ________________________

    ________________________   ________________________

    9. Determine the need for voltage-rated tools and voltage-rated gloves.

☐ ☑ Are voltage-rated tools required?

☐ ☑ Are voltage-rated gloves required?

    10. On 70E-2000 Table 3-3.9.2; 70E-2004 Table 130.7(C)(10) locate the hazard/risk category number. Select the PPE based on the table requirements. Read and comply with all applicable notes to ensure safety.

    PPE required:

Untreated cotton t-shirt	Safety glasses or goggles
Untreated cotton long pants	Double layer switching hood
FR long-sleeved shirt (Note 9)	Hearing protection
FR long pants (Note 9)	Leather gloves
Hard hat	Leather shoes
FR hard hat liner	

    (p. 7 of 10)

TASK ASSESSMENT CHECKLIST FOR 600-VOLT CLASS SWITCHGEAR EQUIPMENT

Job Task A8: Inserting or Removing (Racking) Circuit Breakers from Cubicles with the Doors Closed

Date: _________________________   **Job Location:** _________________________________

Yes   No

☑   ☐   **1.** Is the equipment operating at more that 50 volts or is a shock hazard present?

If "yes," perform a hazard/risk analysis.

If "no," coordinate with the Electrical Safety Program.

☑   ☐   **2.** Is a flash-hazard analysis required? (70E-2000, 2-1.3.3; 70E-2004, 130.3)

3. Determine the shock protection boundary. (70E-2000 Table 2-1.3.4; 70E-2004 Table 130.2(B))
The restricted approach boundary distance is 1 ft, 0 in.

4. Determine the flash protection boundary. (70E-2000, 2-1.3.3.2; 70E-2004, 130.3(A)(1))
The flash protection boundary distance is 4 ft, 0 in.

5. Select the hazard/risk category for the task from Appendix G.
The hazard risk category is 2.

6. Check notes on the table in Appendix G to determine the incident energy level for the specific task in the table.

☑   ☐   **7.** Is the fault energy level available at the location equal to or less than the table notes in Appendix G?
Note: *The available fault current must be 65 kA or less, with 2-cycle clearing time to be applicable.*
If "yes," the table is applicable.   If "no," further investigation is required.

8. Check the legend in Appendix G for additional requirements. List requirements:

________________________________   ________________________________

________________________________   ________________________________

________________________________   ________________________________

9. Determine the need for voltage-rated tools and voltage-rated gloves.

☐   ☑   Are voltage-rated tools required?

☐   ☑   Are voltage-rated gloves required?

10. On 70E-2000 Table 3-3.9.2; 70E-2004 Table 130.7(C)(10) locate the hazard/risk category number. Select the PPE based on the table requirements. Read and comply with all applicable notes to ensure safety.

PPE required:

Untreated cotton t-shirt	Hard hat
Untreated cotton long pants	Safety glasses or goggles
FR long-sleeved shirt	Leather gloves
FR long pants (Note 6)	Leather shoes
Coveralls (Note 7)	

                                                    (p. 8 of 10)

TASK ASSESSMENT CHECKLIST FOR 600-VOLT CLASS SWITCHGEAR EQUIPMENT
Job Task A9: Opening Hinged Covers (to Expose Bare, Energized Parts)

Date: _________________________   **Job Location:** ____

Yes   *No*

☑   ☐   **1.** Is the equipment operating at more that 50 volts or is a shock hazard present?
        If "yes," perform a hazard/risk analysis.
        If "no," coordinate with the Electrical Safety Program.

☑   ☐   **2.** Is a flash-hazard analysis required? (70E-2000, 2-1.3.3; 70E-2004, 130.3)

        3. Determine the shock protection boundary. (70E-2000 Table 2-1.3.4; 70E-2004 Table 130.2(B))
        The restricted approach boundary distance is 1 ft, 0 in.

        4. Determine the flash protection boundary. (70E-2000, 2-1.3.3.2; 70E-2004, 130.3(A)(1))
        The flash protection boundary distance is 4 ft, 0 in.

        5. Select the hazard/risk category for the task from Appendix G.
        The hazard risk category is 2.

        6. Check notes on the table in Appendix G to determine the incident energy level for the specific task in the table.

☑   ☐   **7.** Is the fault energy level available at the location equal to or less than the table notes in Appendix G?
        Note: The available fault current must be 65 kA or less, with 2-cycle clearing time to be applicable.
        If "yes," the table is applicable.   If "no," further investigation is required.

        8. Check the legend in Appendix G for additional requirements. List requirements:

        _______________________________   _______________________________

        _______________________________   _______________________________

        _______________________________   _______________________________

        9. Determine the need for voltage-rated tools and voltage-rated gloves.

☐   ☑   Are voltage-rated tools required?

☐   ☑   Are voltage-rated gloves required?

        10. On 70E-2000 Table 3-3.9.2; 70E-2004 Table 130.7(C)(10) locate the hazard/risk category number. Select the PPE based on the table requirements. Read and comply with all applicable notes to ensure safety.

        PPE required:

        Untreated cotton t-shirt   Hard hat

        Untreated cotton long pants (Note 6)   Safety glasses or goggles

        FR long-sleeved shirt   Leather gloves

        FR long pants (Note 6)   Leather shoes

        Coveralls (Note 7)

                                    (p. 9 of 10)

TASK ASSESSMENT CHECKLIST FOR 600-VOLT CLASS SWITCHGEAR EQUIPMENT
Job Task A10: Applying Safety Grounds, after Voltage Testing

Date: ________________________   **Job Location:** ________________________________

Yes   No

☑   ☐   **1.** Is the equipment operating at more that 50 volts or is a shock hazard present?

If "yes," perform a hazard/risk analysis.

If "no," coordinate with the Electrical Safety Program.

☑   ☐   **2.** Is a flash-hazard analysis required? (70E-2000, 2-1.3.3; 70E-2004, 130.3)

3. Determine the shock protection boundary. (70E-2000 Table 2-1.3.4; 70E-2004 Table 130.2(B))
The restricted approach boundary distance is 1 ft, 0 in.

4. Determine the flash protection boundary. (70E-2000, 2-1.3.3.2; 70E-2004, 130.3(A)(1))
The flash protection boundary distance is 4 ft, 0 in.

5. Select the hazard/risk category for the task from Appendix G.
The hazard risk category is 2.

6. Check notes on the table in Appendix G to determine the incident energy level for the specific task in the table.

☑   ☐   **7.** Is the fault energy level available at the location equal to or less than the table notes in Appendix G?
Note: The available fault current must be 65 kA or less, with 2-cycle clearing time to be applicable.
If "yes," the table is applicable.   If "no," further investigation is required.

8. Check the legend in Appendix G for additional requirements. List requirements:

_______________________   _______________________

_______________________   _______________________

____

9. Determine the need for voltage-rated tools and voltage-rated gloves.

☑   ☐   Are voltage-rated tools required?

☑   ☐   Are voltage-rated gloves required?

10. On 70E-2000 Table 3-3.9.2; 70E-2004 Table 130.7(C)(10) locate the hazard/risk category number. Select the PPE based on the table requirements. Read and comply with all applicable notes to ensure safety.

PPE required:

Untreated cotton t-shirt	Safety glasses or goggles
Untreated cotton long pants (Note 6)	Double layer switching hood
FR long-sleeved shirt (Note 9)	Hearing protection
FR long pants (Note 6)	Leather gloves
Coveralls (Note 7)	Leather shoes

                                    (p. 10 of 10)

TASK ASSESSMENT CHECKLIST FOR 600-VOLT MOTOR CONTROL CENTERS
Job Task B1: Operating a Circuit Breaker or Fused Switch with the Enclosure Door Closed

Date: ________________________   **Job Location:** ____

Yes   No

☑ ☐   **1.** Is the equipment operating at more that 50 volts or is a shock hazard present?

If "yes," perform a hazard/risk analysis.

If "no," coordinate with the Electrical Safety Program.

☑ ☐   **2.** Is a flash-hazard analysis required? (70E-2000, 2-1.3.3; 70E-2004, 130.3)

3. Determine the shock protection boundary. (70E-2000 Table 2-1.3.4; 70E-2004 Table 130.2(B))

The restricted approach boundary distance is 1 ft, 0 in.

4. Determine the flash protection boundary. (70E-2000, 2-1.3.3.2; 70E-2004, 130.3(A)(1))

The flash protection boundary distance is 4 ft, 0 in.

5. Select the hazard/risk category for the task from Appendix G.

The hazard risk category is 0.

6. Check notes on the table in Appendix G to determine the incident energy level for the specific task in the table.

☑ ☐   **7.** Is the fault energy level available at the location equal to or less than the table notes in Appendix G?

Note: The available fault current must be 65 kA or less, with 2-cycle clearing time to be applicable.

If "yes," the table is applicable.   If "no," further investigation is required.

8. Check the legend in Appendix G for additional requirements. List requirements:

________________________________   ________________________________

________________________________   ________________________________

________________________________   ________________________________

9. Determine the need for voltage-rated tools and voltage-rated gloves.

☐ ☑   Are voltage-rated tools required?

☐ ☑   Are voltage-rated gloves required?

10. On 70E-2000 Table 3-3.9.2; 70E-2004 Table 130.7(C)(10) locate the hazard/risk category number. Select the PPE based on the table requirements. Read and comply with all applicable notes to ensure safety.

PPE required:

Untreated cotton long-sleeved shirt

Untreated cotton long pants

Safety glasses

TASK ASSESSMENT CHECKLIST FOR 600-VOLT MOTOR CONTROL CENTERS
Job Task B2: Reading a Panel Meter While Operating a Meter Switch

Date: _______________________   **Job Location:** _______________________________________

Yes   No

☑   ☐   **1.** Is the equipment operating at more that 50 volts or is a shock hazard present?

If "yes," perform a hazard/risk analysis.

If "no," coordinate with the Electrical Safety Program.

☑   ☐   **2.** Is a flash-hazard analysis required? (70E-2000, 2-1.3.3; 70E-2004, 130.3)

3. Determine the shock protection boundary. (70E-2000 Table 2-1.3.4; 70E-2004 Table 130.2(B))

The restricted approach boundary distance is 1 ft, 0 in.

4. Determine the flash protection boundary. (70E-2000, 2-1.3.3.2; 70E-2004, 130.3(A)(1))

The flash protection boundary distance is 4 ft, 0 in.

5. Select the hazard/risk category for the task from Appendix G.

The hazard risk category is 0.

6. Check notes on the table in Appendix G to determine the incident energy level for the specific task in the table.

☑   ☐   **7.** Is the fault energy level available at the location equal to or less than the table notes in Appendix G?

Note: *The available fault current must be 65 kA or less, with 2-cycle clearing time to be applicable.*

If "yes," the table is applicable.   If "no," further investigation is required.

8. Check the legend in Appendix G for additional requirements. List requirements:

___________________________   ___________________________

___________________________   ___________________________

___________________________   ___________________________

9. Determine the need for voltage-rated tools and voltage-rated gloves.

☐   ☑   Are voltage-rated tools required?

☐   ☑   Are voltage-rated gloves required?

10. On 70E-2000 Table 3-3.9.2; 70E-2004 Table 130.7(C)(10) locate the hazard/risk category number. Select the PPE based on the table requirements. Read and comply with all applicable notes to ensure safety.

PPE required:

Untreated cotton long-sleeved shirt

Untreated cotton long pants

Safety glasses

                                                     (p. 2 of 10)

TASK ASSESSMENT CHECKLIST FOR 600-VOLT MOTOR CONTROL CENTERS
Job Task B3: Operating a Circuit Breaker, Fused Switch, or Starter with the Enclosure Door Open

Date: ________________________   **Job Location:** ________________________________

Yes   No

☑ ☐   **1.** Is the equipment operating at more that 50 volts or is a shock hazard present?

   If "yes," perform a hazard/risk analysis.

   If "no," coordinate with the Electrical Safety Program.

☑ ☐   **2.** Is a flash-hazard analysis required? (70E-2000, 2-1.3.3; 70E-2004, 130.3)

   3. Determine the shock protection boundary. (70E-2000 Table 2-1.3.4; 70E-2004 Table 130.2(B))

   The restricted approach boundary distance is 1 ft, 0 in.

   4. Determine the flash protection boundary. (70E-2000, 2-1.3.3.2; 70E-2004, 130.3(A)(1))

   The flash protection boundary distance is 4 ft, 0 in.

   5. Select the hazard/risk category for the task from Appendix G.

   The hazard risk category is 1.

   6. Check notes on the table in Appendix G to determine the incident energy level for the specific task in the table.

☑ ☐   **7.** Is the fault energy level available at the location equal to or less than the table notes in Appendix G?

   Note: *The available fault current must be 65 kA or less, with 2-cycle clearing time to be applicable.*

   If "yes," the table is applicable.   If "no," further investigation is required.

   8. Check the legend in Appendix G for additional requirements. List requirements:

   ______________________________   ______________________________

   ______________________________   ______________________________

   ______________________________   ______________________________

   9. Determine the need for voltage-rated tools and voltage-rated gloves.

☐ ☑   Are voltage-rated tools required?

☐ ☑   Are voltage-rated gloves required?

   10. On 70E-2000 Table 3-3.9.2; 70E-2004 Table 130.7(C)(10) locate the hazard/risk category number. Select the PPE based on the table requirements. Read and comply with all applicable notes to ensure safety.

   PPE required:

   Untreated cotton long pants (Note 4)   Hard hat

   FR long-sleeved shirt   Leather gloves, as needed

   FR pants (Note 4)   Leather shoes, as needed

   Coveralls (Note 5)

   Safety glasses

                                      (p. 3 of 10)

TASK ASSESSMENT CHECKLIST FOR 600-VOLT MOTOR CONTROL CENTERS

Job Task B4: Inserting or Removing Individual Units (Buckets) from a Motor Control Center

Date: _________________________   **Job Location:** _____________________________________

Yes   No

☑ ☐ **1.** Is the equipment operating at more that 50 volts or is a shock hazard present?

   If "yes," perform a hazard/risk analysis.

   If "no," coordinate with the Electrical Safety Program.

☑ ☐ **2.** Is a flash-hazard analysis required? (70E-2000, 2-1.3.3; 70E-2004, 130.3)

   3. Determine the shock protection boundary. (g)

   The restricted approach boundary distance is 1 ft, 0 in.

   4. Determine the flash protection boundary. (70E-2000, 2-1.3.3.2; 70E-2004, 130.3(A)(1))

   The flash protection boundary distance is 4 ft, 0 in.

   5. Select the hazard/risk category for the task from Appendix G.

   The hazard risk category is 3.

   6. Check notes on the table in Appendix G to determine the incident energy level for the specific task in the table.

☑ ☐ **7.** Is the fault energy level available at the location equal to or less than the table notes in Appendix G?

   Note: *The available fault current must be 65 kA or less, with 2-cycle clearing time to be applicable.*

   If "yes," the table is applicable.   If "no," further investigation is required.

   8. Check the legend in Appendix G for additional requirements. List requirements:

   ____________________________   ____________________________

   ____________________________   ____________________________

   ____________________________   ____________________________

   9. Determine the need for voltage-rated tools and voltage-rated gloves.

☑ ☐ Are voltage-rated tools required?

☑ ☐ Are voltage-rated gloves required?

   10. On 70E-2000 Table 3-3.9.2; 70E-2004 Table 130.7(C)(10) locate the hazard/risk category number. Select the PPE based on the table requirements. Read and comply with all applicable notes to ensure safety.

   PPE required:

Untreated cotton t-shirt	Safety glasses or goggles
Untreated cotton long pants	Double layer switching hood
FR long-sleeved shirt (Note 9)	Hearing protection
FR long pants (Note 9)	Leather gloves
Hard hat	Leather shoes
FR hard hat liner	

TASK ASSESSMENT CHECKLIST FOR 600-VOLT CLASS MOTOR CONTROL CENTERS
Job Task B5: Working on Energized Parts, Including Voltage Testing

Date: _________________________   **Job Location:** _____________________________________

Yes   *No*

☑   ☐   **1.** Is the equipment operating at more that 50 volts or is a shock hazard present?

If "yes," perform a hazard/risk analysis.

If "no," coordinate with the Electrical Safety Program.

☑   ☐   **2.** Is a flash-hazard analysis required? (70E-2000, 2-1.3.3; 70E-2004, 130.3)

3. Determine the shock protection boundary. (70E-2000 Table 2-1.3.4; 70E-2004 Table 130.2(B))

The restricted approach boundary distance is 1 ft, 0 in.

4. Determine the flash protection boundary. (70E-2000, 2-1.3.3.2; 70E-2004, 130.3(A)(1))

The flash protection boundary distance is 4 ft, 0 in.

5. Select the hazard/risk category for the task from Appendix G.

The hazard risk category is 2.*

6. Check notes on the table in Appendix G to determine the incident energy level for the specific task in the table.

☑   ☐   **7.** Is the fault energy level available at the location equal to or less than the table notes in Appendix G?

Note: *The available fault current must be 65 kA or less, with 2-cycle clearing time to be applicable.*

If "yes," the table is applicable.   If "no," further investigation is required.

8. Check the legend in Appendix G for additional requirements. List requirements:

__________________________   __________________________

__________________________   __________________________

__________________________   __________________________

9. Determine the need for voltage-rated tools and voltage-rated gloves.

☑   ☐   Are voltage-rated tools required?

☑   ☐   Are voltage-rated gloves required?

10. On 70E-2000 Table 3-3.9.2; 70E-2004 Table 130.7(C)(10) locate the hazard/risk category number. Select the PPE based on the table requirements. Read and comply with all applicable notes to ensure safety.

PPE required:

Untreated cotton t-shirt	Safety glasses or goggles
Untreated cotton long pants (Note 6)	Double layer switching hood
FR long-sleeved shirt (Note 9)	Hearing protection
FR long pants (Note 6)	Leather gloves
Coveralls (Note 7)	Leather shoes
Hard hat	

TASK ASSESSMENT CHECKLIST FOR 600-VOLT CLASS MOTOR CONTROL CENTERS
Job Task B6: Working on Control Circuits with Exposed Parts Energized at 120 Volts or Below

Date: _________________________ **Job Location:** _______________________________________

Yes   No

☑   ☐   **1.** Is the equipment operating at more that 50 volts or is a shock hazard present?

   If "yes," perform a hazard/risk analysis.

   If "no," coordinate with the Electrical Safety Program.

☑   ☐   **2.** Is a flash-hazard analysis required? (70E-2000, 2-1.3.3; 70E-2004, 130.3)

   3. Determine the shock protection boundary. (70E-2000 Table 2-1.3.4; 70E-2004 Table 130.2(B))
   The restricted approach boundary distance is 1 ft, 0 in.

   4. Determine the flash protection boundary. (70E-2000, 2-1.3.3.2; 70E-2004, 130.3(A)(1))
   The flash protection boundary distance is 4 ft, 0 in.

   5. Select the hazard/risk category for the task from Appendix G.
   The hazard risk category is 0.

   6. Check notes on the table in Appendix G to determine the incident energy level for the specific task in the table.

☑   ☐   **7.** Is the fault energy level available at the location equal to or less than the table notes in Appendix G?

   Note: *The available fault current must be 65 kA or less, with 2-cycle clearing time to be applicable.*

   If "yes," the table is applicable.   If "no," further investigation is required.

   8. Check the legend in Appendix G for additional requirements. List requirements:

   _________________________________   _________________________________

   _________________________________   _________________________________

   _________________________________   _________________________________

   9. Determine the need for voltage-rated tools and voltage-rated gloves.

☑   ☐   Are voltage-rated tools required?

☑   ☐   Are voltage-rated gloves required?

   10. On 70E-2000 Table 3-3.9.2; 70E-2004 Table 130.7(C)(10) locate the hazard/risk category number. Select the PPE based on the table requirements. Read and comply with all applicable notes to ensure safety.

   PPE required:
   Untreated cotton long-sleeved t-shirt
   Untreated cotton long pants
   Safety glasses

                                   (p. 6 of 10)

TASK ASSESSMENT CHECKLIST FOR 600-VOLT MOTOR CONTROL CENTERS
Job Task B7: Working on Control Circuits with Exposed Parts Energized at More Than 120 Volts

Date: ________________________   **Job Location:** ____

Yes   No

☑   ☐   **1.** Is the equipment operating at more that 50 volts or is a shock hazard present?

If "yes," perform a hazard/risk analysis.

If "no," coordinate with the Electrical Safety Program.

☑   ☐   **2.** Is a flash-hazard analysis required? (70E-2000, 2-1.3.3; 70E-2004, 130.3)

3. Determine the shock protection boundary. (70E-2000 Table 2-1.3.4; 70E-2004 Table 130.2(B))

The restricted approach boundary distance is 1 ft, 0 in.

4. Determine the flash protection boundary. (70E-2000, 2-1.3.3.2; 70E-2004, 130.3(A)(1))

The flash protection boundary distance is 4 ft, 0 in.

5. Select the hazard/risk category for the task from Appendix G.

The hazard risk category is 2.*

6. Check notes on the table in Appendix G to determine the incident energy level for the specific task in the table.

☑   ☐   **7.** Is the fault energy level available at the location equal to or less than the table notes in Appendix G?

Note: The available fault current must be 65 kA or less, with 2-cycle clearing time to be applicable.

If "yes," the table is applicable.   If "no," further investigation is required.

8. Check the legend in Appendix G for additional requirements. List requirements:

_______________________________   ____

_______________________________   ____

_______________________________   ____

9. Determine the need for voltage-rated tools and voltage-rated gloves.

☑   ☐   Are voltage-rated tools required?

☑   ☐   Are voltage-rated gloves required?

10. On 70E-2000 Table 3-3.9.2; 70E-2004 Table 130.7(C)(10) locate the hazard/risk category number. Select the PPE based on the table requirements. Read and comply with all applicable notes to ensure safety.

PPE required:

Untreated cotton t-shirt	Safety glasses or goggles
Untreated cotton long pants (Note 6)	Double layer switching hood
FR long-sleeved shirt (Note 9)	Hearing protection
FR long pants (Note 6)	Leather gloves
Coveralls (Note 7)	Leather shoes
Hard hat	

                    (p. 7 of 10)

TASK ASSESSMENT CHECKLIST FOR 600-VOLT MOTOR CONTROL CENTERS
Job Task B8: Removing Bolted Covers (to Expose Bare, Energized Parts)

Date: _________________________   **Job Location:** _________________________________

Yes   No

☑   ☐   **1.** Is the equipment operating at more that 50 volts or is a shock hazard present?

If "yes," perform a hazard/risk analysis.

If "no," coordinate with the Electrical Safety Program.

☑   ☐   **2.** Is a flash-hazard analysis required? (70E-2000, 2-1.3.3; 70E-2004, 130.3)

3. Determine the shock protection boundary. (70E-2000 Table 2-1.3.4; 70E-2004 Table 130.2(B))

The restricted approach boundary distance is 1 ft, 0 in.

4. Determine the flash protection boundary. (70E-2000, 2-1.3.3.2; 70E-2004, 130.3(A)(1))

The flash protection boundary distance is 4 ft, 0 in.

5. Select the hazard/risk category for the task from Appendix G.

The hazard risk category is 2.*

6. Check notes on the table in Appendix G to determine the incident energy level for the specific task in the table.

☑   ☐   **7.** Is the fault energy level available at the location equal to or less than the table notes in Appendix G?

Note: *The available fault current must be 65 kA or less, with 2-cycle clearing time to be applicable.*

If "yes," the table is applicable.   If "no," further investigation is required.

8. Check the legend in Appendix G for additional requirements. List requirements:

_________________________________   _________________________________

_________________________________   _________________________________

____

9. Determine the need for voltage-rated tools and voltage-rated gloves.

☑   ☐   Are voltage-rated tools required?

☑   ☐   Are voltage-rated gloves required?

10. On 70E-2000 Table 3-3.9.2; 70E-2004 Table 130.7(C)(10) locate the hazard/risk category number. Select the PPE based on the table requirements. Read and comply with all applicable notes to ensure safety.

PPE required:

Untreated cotton t-shirt	Safety glasses or goggles
Untreated cotton long pants (Note 6)	Double layer switching hood
FR long-sleeved shirt (Note 9)	Hearing protection
FR long pants (Note 6)	Leather gloves
Coveralls (Note 7)	Leather shoes
Hard hat	

TASK ASSESSMENT CHECKLIST FOR 600-VOLT MOTOR CONTROL CENTERS
Job Task B9: Opening Hinged Covers (to Expose Bare, Energized Parts)

Date: _________________________ Job Location: _____________________________________

Yes No

[✓] [ ] 1. Is the equipment operating at more that 50 volts or is a shock hazard present?

If "yes," perform a hazard/risk analysis.

If "no," coordinate with the Electrical Safety Program.

[✓] [ ] 2. Is a flash-hazard analysis required? (70E-2000, 2-1.3.3; 70E-2004, 130.3)

3. Determine the shock protection boundary. (70E-2000 Table 2-1.3.4; 70E-2004 Table 130.2(B))

The restricted approach boundary distance is 1 ft, 0 in.

4. Determine the flash protection boundary. (70E-2000, 2-1.3.3.2; 70E-2004, 130.3(A)(1))

The flash protection boundary distance is 4 ft, 0 in.

5. Select the hazard/risk category for the task from Appendix G.

The hazard risk category is 1.

6. Check notes on the table in Appendix G to determine the incident energy level for the specific task in the table.

[✓] [ ] 7. Is the fault energy level available at the location equal to or less than the table notes in Appendix G?

Note: *The available fault current must be 65 kA or less, with 2-cycle clearing time to be applicable.*

If "yes," the table is applicable.   If "no," further investigation is required.

8. Check the legend in Appendix G for additional requirements. List requirements:

_________________________________   _________________________________

_________________________________   _________________________________

_________________________________   _________________________________

9. Determine the need for voltage-rated tools and voltage-rated gloves.

[ ] [✓] Are voltage-rated tools required?

[ ] [✓] Are voltage-rated gloves required?

10. On 70E-2000 Table 3-3.9.2; 70E-2004 Table 130.7(C)(10) locate the hazard/risk category number. Select the PPE based on the table requirements. Read and comply with all applicable notes to ensure safety.

PPE required:

Untreated cotton long pants (Note 4)	Hard hat
FR long-sleeved shirt	Safety glasses
FR pants (Note 4)	Leather gloves
Coveralls (Note 5)	Leather shoes, as needed

   (p. 9 of 10)

TASK ASSESSMENT CHECKLIST FOR 600-VOLT MOTOR CONTROL CENTERS
Job Task B10: Applying Safety Grounds after Voltage Testing

Date: _________________________   **Job Location:** _________________________________

Yes   No

☑ ☐   **1.** Is the equipment operating at more that 50 volts or is a shock hazard present?

If "yes," perform a hazard/risk analysis.

If "no," coordinate with the Electrical Safety Program.

☑ ☐   **2.** Is a flash-hazard analysis required? (70E-2000, 2-1.3.3; 70E-2004, 130.3)

3. Determine the shock protection boundary. (70E-2000 Table 2-1.3.4; 70E-2004 Table 130.2(B))

The restricted approach boundary distance is 1 ft, 0 in.

4. Determine the flash protection boundary. (70E-2000, 2-1.3.3.2; 70E-2004, 130.3(A)(1))

The flash protection boundary distance is 4 ft, 0 in.

5. Select the hazard/risk category for the task from Appendix G.

The hazard risk category is 2.*

6. Check notes on the table in Appendix G to determine the incident energy level for the specific task in the table.

☑ ☐   **7.** Is the fault energy level available at the location equal to or less than the table notes in Appendix G?

Note: *The available fault current must be 65 kA or less, with 2-cycle clearing time to be applicable.*

If "yes," the table is applicable.   If "no," further investigation is required.

8. Check the legend in Appendix G for additional requirements. List requirements:

_________________________________   _________________________________

_________________________________   _________________________________

_________________________________   _________________________________

9. Determine the need for voltage-rated tools and voltage-rated gloves.

☑ ☐   Are voltage-rated tools required?

☑ ☐   Are voltage-rated gloves required?

10. On 70E-2000 Table 3-3.9.2; 70E-2004 Table 130.7(C)(10) locate the hazard/risk category number. Select the PPE based on the table requirements. Read and comply with all applicable notes to ensure safety.

PPE required:

Untreated cotton t-shirt	Safety glasses or goggles
Untreated cotton long pants (Note 6)	Double layer switching hood
FR long-sleeved shirt (Note 9)	Hearing protection
FR long pants (Note 6)	Leather gloves
Coveralls (Note 7)	Leather shoes
Hard hat	

                    (p. 10 of 10)

TASK ASSESSMENT CHECKLIST FOR PANELBOARDS RATED
MORE THAN 240 VOLTS UP TO 600 VOLTS

Job Task C1: Operating a Circuit Breaker or Fused Switch with the Cover On

Date: _________________________   **Job Location:** ____

Yes   *No*

☑   ☐   **1.** Is the equipment operating at more that 50 volts or is a shock hazard present?

           If "yes," perform a hazard/risk analysis.

           If "no," coordinate with the Electrical Safety Program.

☑   ☐   **2.** Is a flash-hazard analysis required? (70E-2000, 2-1.3.3; 70E-2004, 130.3)

           3. Determine the shock protection boundary. (70E-2000 Table 2-1.3.4; 70E-2004 Table 130.2(B))

           The restricted approach boundary distance is 1 ft, 0 in.

           4. Determine the flash protection boundary. (70E-2000, 2-1.3.3.2; 70E-2004, 130.3(A)(1))

           The flash protection boundary distance is 4 ft, 0 in.

           5. Select the hazard/risk category for the task from Appendix G.

           The hazard risk category is 0.

           6. Check notes on the table in Appendix G to determine the incident energy level for the specific task in the table.

☑   ☐   **7.** Is the fault energy level available at the location equal to or less than the table notes in Appendix G?

           Note: *The available fault current must be 25 kA or less, with 2-cycle clearing time to be applicable.*

           If "yes," the table is applicable.   If "no," further investigation is required.

           8. Check the legend in Appendix G for additional requirements. List requirements:

           ________________________   ________________________

           ________________________   ________________________

           ________________________   ________________________

           9. Determine the need for voltage-rated tools and voltage-rated gloves.

☐   ☑   Are voltage-rated tools required?

☐   ☑   Are voltage-rated gloves required?

           10. On 70E-2000 Table 3-3.9.2; 70E-2004 Table 130.7(C)(10) locate the hazard/risk category number. Select the PPE based on the table requirements. Read and comply with all applicable notes to ensure safety.

           PPE required:

           Untreated cotton long-sleeved shirt

           Untreated cotton long pants

           Safety glasses

        (p. 1 of 3)

TASK ASSESSMENT CHECKLIST FOR PANELBOARDS RATED MORE THAN 240 VOLTS UP TO 600 VOLTS

Job Task C2: Operating a Circuit Breaker or Fused Switch with the Cover Off

Date: _______________________   **Job Location:** _______________________________________

Yes	**No**	
☑	☐	**1.** Is the equipment operating at more that 50 volts or is a shock hazard present?

If "yes," perform a hazard/risk analysis.

If "no," coordinate with the Electrical Safety Program.

☑ ☐ **2.** Is a flash-hazard analysis required? (70E-2000, 2-1.3.3; 70E-2004, 130.3)

3. Determine the shock protection boundary. (70E-2000 Table 2-1.3.4; 70E-2004 Table 130.2(B))

The restricted approach boundary distance is 1 ft, 0 in.

4. Determine the flash protection boundary. (70E-2000, 2-1.3.3.2; 70E-2004, 130.3(A)(1))

The flash protection boundary distance is 4 ft, 0 in.

5. Select the hazard/risk category for the task from Appendix G.

The hazard risk category is 1.

6. Check notes on the table in Appendix G to determine the incident energy level for the specific task in the table.

☑ ☐ **7.** Is the fault energy level available at the location equal to or less than the table notes in Appendix G?

Note: The available fault current must be 25 kA or less, with 2-cycle clearing time to be applicable.

If "yes," the table is applicable.   If "no," further investigation is required.

8. Check the legend in Appendix G for additional requirements. List requirements:

_______________________________   _______________________________

_______________________________   _______________________________

_______________________________   _______________________________

9. Determine the need for voltage-rated tools and voltage-rated gloves.

☐ ☑   Are voltage-rated tools required?

☐ ☑   Are voltage-rated gloves required?

10. On 70E-2000 Table 3-3.9.2; 70E-2004 Table 130.7(C)(10) locate the hazard/risk category number. Select the PPE based on the table requirements. Read and comply with all applicable notes to ensure safety.

PPE required:

Untreated cotton long pants (Note 4)	Hard hat
FR long-sleeved shirt	Safety glasses
FR pants (Note 4)	Leather gloves, as needed
Coveralls (Note 5)	Leather shoes, as needed

TASK ASSESSMENT CHECKLIST FOR PANELBOARDS RATED MORE THAN 240 VOLTS UP TO 600 VOLTS

Job Task C3: Working on Energized Parts, Including Voltage Testing

Date: _________________________   **Job Location:** ____

Yes *No*

☑ ☐ **1.** Is the equipment operating at more that 50 volts or is a shock hazard present?

If "yes," perform a hazard/risk analysis.

If "no," coordinate with the Electrical Safety Program.

☑ ☐ **2.** Is a flash-hazard analysis required? (70E-2000, 2-1.3.3; 70E-2004, 130.3)

3. Determine the shock protection boundary. (70E-2000 Table 2-1.3.4; 70E-2004 Table 130.2(B))

The restricted approach boundary distance is 1 ft, 0 in.

4. Determine the flash protection boundary. (70E-2000, 2-1.3.3.2; 70E-2004, 130.3(A)(1))

The flash protection boundary distance is 4 ft, 0 in.

5. Select the hazard/risk category for the task from Appendix G.

The hazard risk category is 2*.

6. Check notes on the table in Appendix G to determine the incident energy level for the specific task in the table.

☑ ☐ **7.** Is the fault energy level available at the location equal to or less than the table notes in Appendix G?

Note: *The available fault current must be 25 kA or less, with 2-cycle clearing time to be applicable.*

If "yes," the table is applicable.   If "no," further investigation is required.

8. Check the legend in Appendix G for additional requirements. List requirements:

_________________________________   _________________________________

_________________________________   _________________________________

_________________________________   _________________________________

9. Determine the need for voltage-rated tools and voltage-rated gloves.

☑ ☐ Are voltage-rated tools required?

☑ ☐ Are voltage-rated gloves required?

10. On 70E-2000 Table 3-3.9.2; 70E-2004 Table 130.7(C)(10) locate the hazard/risk category number. Select the PPE based on the table requirements. Read and comply with all applicable notes to ensure safety.

PPE required:

Untreated cotton t-shirt	Safety glasses or goggles
Untreated cotton long pants (Note 6)	Double layer switching hood
FR long-sleeved shirt (Note 9)	Hearing protection
FR long pants (Note 6)	Leather gloves
Coveralls (Note 7)	Leather shoes
Hard hat	

TASK ASSESSMENT CHECKLIST FOR PANELBOARDS RATED 240 VOLTS AND BELOW
Job Task D1: Operating a Circuit Breaker or Fused Switch with the Cover On

Date: _______________________   Job Location: _____________________________________

Yes   No

☑   ☐   **1.** Is the equipment operating at more that 50 volts or is a shock hazard present?

If "yes," perform a hazard/risk analysis.

If "no," coordinate with the Electrical Safety Program.

☑   ☐   **2.** Is a flash-hazard analysis required? (70E-2000, 2-1.3.3; 70E-2004, 130.3)

3. Determine the shock protection boundary. (70E-2000 Table 2-1.3.4; 70E-2004 Table 130.2(B))

The restricted approach boundary distance is "avoid contact."

4. Determine the flash protection boundary. (70E-2000, 2-1.3.3.2; 70E-2004, 130.3(A)(1))

The flash protection boundary distance is 4 ft, 0 in.

5. Select the hazard/risk category for the task from Appendix G.

The hazard risk category is 0.

6. Check notes on the table in Appendix G to determine the incident energy level for the specific task in the table.

☑   ☐   **7.** Is the fault energy level available at the location equal to or less than the table notes in Appendix G?

Note: *The available fault current must be 25 kA or less, with 2-cycle clearing time to be applicable.*

If "yes," the table is applicable.   If "no," further investigation is required.

8. Check the legend in Appendix G for additional requirements. List requirements:

_____________________________   _____________________________________

_____________________________   _____________________________________

_____________________________   _____________________________________

9. Determine the need for voltage-rated tools and voltage-rated gloves.

☐   ☑   Are voltage-rated tools required?

☐   ☑   Are voltage-rated gloves required?

10. On 70E-2000 Table 3-3.9.2; 70E-2004 Table 130.7(C)(10) locate the hazard/risk category number. Select the PPE based on the table requirements. Read and comply with all applicable notes to ensure safety.

PPE required:

Untreated cotton long-sleeved shirt

Untreated cotton long pants

Safety glasses

TASK ASSESSMENT CHECKLIST FOR PANELBOARDS RATED 240 VOLTS AND BELOW
Job Task D2: Operating a Circuit Breaker or Fused Switch with the Cover Off

Date: ________________________   **Job Location:** ________________________________

Yes   No

☑   ☐   **1.** Is the equipment operating at more that 50 volts or is a shock hazard present?

If "yes," perform a hazard/risk analysis.

If "no," coordinate with the Electrical Safety Program.

☑   ☐   **2.** Is a flash-hazard analysis required? (70E-2000, 2-1.3.3; 70E-2004, 130.3)

3. Determine the shock protection boundary. (70E-2000 Table 2-1.3.4; 70E-2004 Table 130.2(B))

The restricted approach boundary distance is "avoid contact."

4. Determine the flash protection boundary. (70E-2000, 2-1.3.3.2; 70E-2004, 130.3(A)(1))

The flash protection boundary distance is 4 ft, 0 in.

5. Select the hazard/risk category for the task from Appendix G.

The hazard risk category is 0.

6. Check notes on the table in Appendix G to determine the incident energy level for the specific task in the table.

☑   ☐   **7.** Is the fault energy level available at the location equal to or less than the table notes in Appendix G?

Note: *The available fault current must be 25 kA or less, with 2-cycle clearing time to be applicable.*

If "yes," the table is applicable.   If "no," further investigation is required.

8. Check the legend in Appendix G for additional requirements. List requirements:

________________________________   ________________________________

________________________________   ________________________________

________________________________   ________________________________

9. Determine the need for voltage-rated tools and voltage-rated gloves.

☐   ☑   Are voltage-rated tools required?

☐   ☑   Are voltage-rated gloves required?

10. On 70E-2000 Table 3-3.9.2; 70E-2004 Table 130.7(C)(10) locate the hazard/risk category number. Select the PPE based on the table requirements. Read and comply with all applicable notes to ensure safety.

PPE required:

Untreated cotton long-sleeved shirt

Untreated cotton long pants

Safety glasses

TASK ASSESSMENT CHECKLIST FOR PANELBOARDS RATED 240 VOLTS AND BELOW
Job Task D3: Working on Energized Parts, Including Voltage Testing

Date: _________________________   **Job Location:** _______________________________________

Yes   *No*

☑   ☐   **1.** Is the equipment operating at more that 50 volts or is a shock hazard present?

If "yes," perform a hazard/risk analysis.

If "no," coordinate with the Electrical Safety Program.

☑   ☐   **2.** Is a flash-hazard analysis required? (70E-2000, 2-1.3.3; 70E-2004, 130.3)

3. Determine the shock protection boundary. (70E-2000 Table 2-1.3.4; 70E-2004 Table 130.2(B))

The restricted approach boundary distance is "avoid contact."

4. Determine the flash protection boundary. (70E-2000, 2-1.3.3.2; 70E-2004, 130.3(A)(1))

The flash protection boundary distance is 4 ft, 0 in.

5. Select the hazard/risk category for the task from Appendix G.

The hazard risk category is 1.

6. Check notes on the table in Appendix G to determine the incident energy level for the specific task in the table.

☑   ☐   **7.** Is the fault energy level available at the location equal to or less than the table notes in Appendix G?

Note: The available fault current must be 25 kA or less, with 2-cycle clearing time to be applicable.

If "yes," the table is applicable.   If "no," further investigation is required.

8. Check the legend in Appendix G for additional requirements. List requirements:

________________________________   ________________________________

________________________________   ________________________________

________________________________   ________________________________

9. Determine the need for voltage-rated tools and voltage-rated gloves.

☑   ☐   Are voltage-rated tools required?

☑   ☐   Are voltage-rated gloves required?

10. On 70E-2000 Table 3-3.9.2; 70E-2004 Table 130.7(C)(10) locate the hazard/risk category number. Select the PPE based on the table requirements. Read and comply with all applicable notes to ensure safety.

PPE required:

Untreated cotton long pants (Note 4)	Hard hat
FR long-sleeved shirt	Safety glasses
FR pants (Note 4)	Leather gloves, as needed
Coveralls (Note 5)	Leather shoes, as needed

                                        (p. 3 of 6)

TASK ASSESSMENT CHECKLIST FOR PANELBOARDS RATED 240 VOLTS AND BELOW
Job Task D4: Removing/Installing Circuit Breakers or Fused Switches

Date: _______________________   **Job Location:** _______________________________________

Yes   **No**

☑   ☐   **1.** Is the equipment operating at more that 50 volts or is a shock hazard present?

         If "yes," perform a hazard/risk analysis.

         If "no," coordinate with the Electrical Safety Program.

☑   ☐   **2.** Is a flash-hazard analysis required? (70E-2000, 2-1.3.3; 70E-2004, 130.3)

         3. Determine the shock protection boundary. (70E-2000 Table 2-1.3.4; 70E-2004 Table 130.2(B))

         The restricted approach boundary distance is "avoid contact."

         4. Determine the flash protection boundary. (70E-2000, 2-1.3.3.2; 70E-2004, 130.3(A)(1))

         The flash protection boundary distance is 4 ft, 0 in.

         5. Select the hazard/risk category for the task from Appendix G.

         The hazard risk category is 1.

         6. Check notes on the table in Appendix G to determine the incident energy level for the specific task in the table.

☑   ☐   **7.** Is the fault energy level available at the location equal to or less than the table notes in Appendix G?

         Note: *The available fault current must be 25 kA or less, with 2-cycle clearing time to be applicable.*

         If "yes," the table is applicable.   If "no," further investigation is required.

         8. Check the legend in Appendix G for additional requirements. List requirements:

         _______________________________   _______________________________

         _______________________________   _______________________________

         _______________________________   _______________________________

         9. Determine the need for voltage-rated tools and voltage-rated gloves.

☑   ☐   Are voltage-rated tools required?

☑   ☐   Are voltage-rated gloves required?

       10. On 70E-2000 Table 3-3.9.2; 70E-2004 Table 130.7(C)(10) locate the hazard/risk category number. Select the PPE based on the table requirements. Read and comply with all applicable notes to ensure safety.

         PPE required:

Untreated cotton long pants (Note 4)	Hard hat
FR long-sleeved shirt	Safety glasses
FR pants (Note 4)	Leather gloves, as needed
Coveralls (Note 5)	Leather shoes, as needed

      (p. 4 of 6)

TASK ASSESSMENT CHECKLIST FOR PANELBOARDS RATED 240 VOLTS AND BELOW
Job Task D5: Removing Bolted Covers (to Expose Bare, Energized Parts)

Date: ________________________   Job Location: ________________________________

Yes   No

☑   ☐   **1.** Is the equipment operating at more that 50 volts or is a shock hazard present?

If "yes," perform a hazard/risk analysis.

If "no," coordinate with the Electrical Safety Program.

☑   ☐   **2.** Is a flash-hazard analysis required? (70E-2000, 2-1.3.3; 70E-2004, 130.3)

3. Determine the shock protection boundary. (70E-2000 Table 2-1.3.4; 70E-2004 Table 130.2(B))

The restricted approach boundary distance is "avoid contact."

4. Determine the flash protection boundary. (70E-2000, 2-1.3.3.2; 70E-2004, 130.3(A)(1))

The flash protection boundary distance is 4 ft, 0 in.

5. Select the hazard/risk category for the task from Appendix G.

The hazard risk category is 1.

6. Check notes on the table in Appendix G to determine the incident energy level for the specific task in the table.

☑   ☐   **7.** Is the fault energy level available at the location equal to or less than the table notes in Appendix G?

Note: The available fault current must be 25 kA or less, with 2-cycle clearing time to be applicable.

If "yes," the table is applicable.   If "no," further investigation is required.

8. Check the legend in Appendix G for additional requirements. List requirements:

____________________________   ____________________________

____________________________   ____________________________

____________________________   ____________________________

9. Determine the need for voltage-rated tools and voltage-rated gloves.

☑   ☐   Are voltage-rated tools required?

☑   ☐   Are voltage-rated gloves required?

10. On 70E-2000 Table 3-3.9.2; 70E-2004 Table 130.7(C)(10) locate the hazard/risk category number. Select the PPE based on the table requirements. Read and comply with all applicable notes to ensure safety.

PPE required:

Untreated cotton long pants (Note 4)	Hard hat
FR long-sleeved shirt	Safety glasses
FR pants (Note 4)	Leather gloves, as needed
Coveralls (Note 5)	Leather shoes, as needed

TASK ASSESSMENT CHECKLIST FOR PANELBOARDS RATED 240 VOLTS AND BELOW
Job Task D6: Opening Hinged Covers (to Expose Bare, Energized Parts)

Date: _________________________   Job Location: _____________________________

Yes   **No**

☑   ☐   **1.** Is the equipment operating at more that 50 volts or is a shock hazard present?

If "yes," perform a hazard/risk analysis.

If "no," coordinate with the Electrical Safety Program.

☑   ☐   **2.** Is a flash-hazard analysis required? (70E-2000, 2-1.3.3; 70E-2004, 130.3)

3. Determine the shock protection boundary. (70E-2000 Table 2-1.3.4; 70E-2004 Table 130.2(B))

The restricted approach boundary distance is "avoid contact."

4. Determine the flash protection boundary. (70E-2000, 2-1.3.3.2; 70E-2004, 130.3(A)(1))

The flash protection boundary distance is 4 ft, 0 in.

5. Select the hazard/risk category for the task from Appendix G.

The hazard risk category is 0.

6. Check notes on the table in Appendix G to determine the incident energy level for the specific task in the table.

☑   ☐   **7.** Is the fault energy level available at the location equal to or less than the table notes in Appendix G?

Note: *The available fault current must be 25 kA or less, with 2-cycle clearing time to be applicable.*

If "yes," the table is applicable.   If "no," further investigation is required.

8. Check the legend in Appendix G for additional requirements. List requirements:

______________________________   ______________________________

______________________________   ______________________________

______________________________   ______________________________

9. Determine the need for voltage-rated tools and voltage-rated gloves.

☑   ☐   Are voltage-rated tools required?

☑   ☐   Are voltage-rated gloves required?

10. On 70E-2000 Table 3-3.9.2; 70E-2004 Table 130.7(C)(10) locate the hazard/risk category number. Select the PPE based on the table requirements. Read and comply with all applicable notes to ensure safety.

PPE required:

Untreated cotton long-sleeved shirt

Untreated cotton long pants

Safety glasses

**TASK ASSESSMENT CHECKLIST FOR COMMON TASKS
ON SYSTEMS RATED 600 VOLTS AND BELOW**

Job Task E1: Removing/Replacing Light Fixture Ballast (Light Fixture Rated 277 Volts and Below)

Date: _________________________ **Job Location:** ____

Yes   No

☑   ☐   **1.** Is the equipment operating at more that 50 volts or is a shock hazard present?

  If "yes," perform a hazard/risk analysis.

  If "no," coordinate with the Electrical Safety Program.

☑   ☐   **2.** Is a flash-hazard analysis required? (70E-2000, 2-1.3.3; 70E-2004, 130.3)

  3. Determine the shock protection boundary. (70E-2000 Table 2-1.3.4; 70E-2004 Table 130.2(B))

  The restricted approach boundary distance is "avoid contact."

  4. Determine the flash protection boundary. (70E-2000, 2-1.3.3.2; 70E-2004, 130.3(A)(1))

  The flash protection boundary distance is less than 4 ft, 0 in.

  5. Select the hazard/risk category for the task from Appendix G.

  The hazard risk category is 0.

  6. Check notes on the table in Appendix G to determine the incident energy level for the specific task in the table.

☑   ☐   **7.** Is the fault energy level available at the location equal to or less than the table notes in Appendix G?

  Note: *The available fault current must be 25 kA or less, with 2-cycle clearing time to be applicable.*

  If "yes," the table is applicable.   If "no," further instigation is required.

  8. Check the legend in Appendix G for additional requirements. List requirements:

  ________________________________   ________________________________

  ________________________________   ________________________________

  ________________________________   ________________________________

  9. Determine the need for voltage-rated tools and voltage-rated gloves.

☑   ☐   Are voltage-rated tools required?

☑   ☐   Are voltage-rated gloves required?

  10. On 70E-2000 Table 3-3.9.2; 70E-2004 Table 130.7(C)(10) locate the hazard/risk category number. Select the PPE based on the table requirements. Read and comply with all applicable notes to ensure safety.

  PPE required:

  Untreated cotton long-sleeved shirt

  Untreated cotton long pants

  Safety glasses

                                               (p. 1 of 6)

TASK ASSESSMENT CHECKLIST FOR COMMON TASKS
ON SYSTEMS RATED 600 VOLTS AND BELOW

Job Task E2: Replacing 15- and 20-Ampere Receptacle or Switch (Devices Rated 277 Volts and Below)

Date: ________________________   **Job Location:** ____

Yes *No*

☑ ☐ **1.** Is the equipment operating at more that 50 volts or is a shock hazard present?

If "yes," perform a hazard/risk analysis.

If "no," coordinate with the Electrical Safety Program.

☑ ☐ **2.** Is a flash-hazard analysis required? (70E-2000, 2-1.3.3; 70E-2004, 130.3)

3. Determine the shock protection boundary. (70E-2000 Table 2-1.3.4; 70E-2004 Table 130.2(B))

The restricted approach boundary distance is "avoid contact."

4. Determine the flash protection boundary. (70E-2000, 2-1.3.3.2; 70E-2004, 130.3(A)(1))

The flash protection boundary distance is not applicable.

5. Select the hazard/risk category for the task from Appendix G.

The hazard risk category is 0.

6. Check notes on the table in Appendix G to determine the incident energy level for the specific task in the table.

☑ ☐ **7.** Is the fault energy level available at the location equal to or less than the table notes in Appendix G?

Note: *The available fault current must be 25 kA or less, with 2-cycle clearing time to be applicable.*

If "yes," the table is applicable.   If "no," further investigation is required.

8. Check the legend in Appendix G for additional requirements. List requirements:

___________________________   ___________________________

___________________________   ___________________________

___________________________   ___________________________

9. Determine the need for voltage-rated tools and voltage-rated gloves.

☑ ☐ Are voltage-rated tools required?

☑ ☐ Are voltage-rated gloves required?

10. On 70E-2000 Table 3-3.9.2; 70E-2004 Table 130.7(C)(10) locate the hazard/risk category number. Select the PPE based on the table requirements. Read and comply with all applicable notes to ensure safety.

PPE required:

Untreated cotton long-sleeved shirt

Untreated cotton long pants

Safety glasses

TASK ASSESSMENT CHECKLIST FOR COMMON TASKS
ON SYSTEMS RATED 600 VOLTS AND BELOW

Job Task E3: Disconnecting/Reconnecting Utilization Equipment Rated 240 Volts or Below

Date: ___________________   **Job Location:** ____

Yes   No

☑ ☐ **1.** Is the equipment operating at more that 50 volts or is a shock hazard present?

   If "yes," perform a hazard/risk analysis.

   If "no," coordinate with the Electrical Safety Program.

☑ ☐ **2.** Is a flash-hazard analysis required? (70E-2000, 2-1.3.3; 70E-2004, 130.3)

   3. Determine the shock protection boundary. (70E-2000 Table 2-1.3.4; 70E-2004 Table 130.2(B))

   The restricted approach boundary distance is "avoid contact."

   4. Determine the flash protection boundary. (70E-2000, 2-1.3.3.2; 70E-2004, 130.3(A)(1))

   The flash protection boundary distance is 4 ft, 0 in.

   5. Select the hazard/risk category for the task from Appendix G.

   The hazard risk category is 1.

   6. Check notes on the table in Appendix G to determine the incident energy level for the specific task in the table.

☑ ☐ **7.** Is the fault energy level available at the location equal to or less than the table notes in Appendix G?

   Note: *The available fault current must be 25 kA or less, with 2-cycle clearing time to be applicable.*

   If "yes," the table is applicable.   If "no," further investigation is required.

   8. Check the legend in Appendix G for additional requirements. List requirements:

   _______________________________   _______________________________

   _______________________________   _______________________________

   _______________________________   _______________________________

   9. Determine the need for voltage-rated tools and voltage-rated gloves.

☑ ☐ Are voltage-rated tools required?

☑ ☐ Are voltage-rated gloves required?

   10. On 70E-2000 Table 3-3.9.2; 70E-2004 Table 130.7(C)(10) locate the hazard/risk category number. Select the PPE based on the table requirements. Read and comply with all applicable notes to ensure safety.

   PPE required:

Untreated cotton long pants (Note 4)	Hard hat
FR long-sleeved shirt	Safety glasses
FR pants (Note 4)	Leather gloves
Coveralls (Note 5)	Leather shoes, as needed

TASK ASSESSMENT CHECKLIST FOR COMMON TASKS
ON SYSTEMS RATED 600 VOLTS AND BELOW

Job Task E4: Disconnecting/Reconnecting Utilization
Equipment Rated More Than 240 volts up to 600 Volts

Date: ________________________   **Job Location:** ______________________________________

Yes   No

☑ ☐ **1.** Is the equipment operating at more that 50 volts or is a shock hazard present?

   If "yes," perform a hazard/risk analysis.

   If "no," coordinate with the Electrical Safety Program.

☑ ☐ **2.** Is a flash-hazard analysis required? (70E-2000, 2-1.3.3; 70E-2004, 130.3)

3. Determine the shock protection boundary. (70E-2000 Table 2-1.3.4; 70E-2004 Table 130.2(B))
   The restricted approach boundary distance is 1 ft, 0 in.

4. Determine the flash protection boundary. (70E-2000, 2-1.3.3.2; 70E-2004, 130.3(A)(1))
   The flash protection boundary distance is 4 ft, 0 in.

5. Select the hazard/risk category for the task from Appendix G.
   The hazard risk category is 1.

6. Check notes on the table in Appendix G to determine the incident energy level for the specific task in the table.

☑ ☐ **7.** Is the fault energy level available at the location equal to or less than the table notes in Appendix G?

   Note: The available fault current must be 25 kA or less, with 2-cycle clearing time to be applicable.

   If "yes," the table is applicable.   If "no," further investigation is required.

8. Check the legend in Appendix G for additional requirements. List requirements:

   ________________________________   ________________________________

   ________________________________   ________________________________

   ________________________________   ________________________________

9. Determine the need for voltage-rated tools and voltage-rated gloves.

☑ ☐   Are voltage-rated tools required?

☑ ☐   Are voltage-rated gloves required?

10. On 70E-2000 Table 3-3.9.2; 70E-2004 Table 130.7(C)(10) locate the hazard/risk category number. Select the PPE based on the table requirements. Read and comply with all applicable notes to ensure safety.

   PPE required:

Untreated cotton long pants (Note 4)	Hard hat
FR long-sleeved shirt	Safety glasses
FR pants (Note 4)	Leather gloves

                                        (p. 4 of 6)

TASK ASSESSMENT CHECKLIST FOR COMMON TASKS ON SYSTEMS RATED 600 VOLTS AND BELOW

Job Task E5: Voltage Testing at Utilization Equipment Rated 240 Volts or Below

Date: _______________________   **Job Location:** ____

Yes *No*

☑ ☐ **1.** Is the equipment operating at more that 50 volts or is a shock hazard present?

        If "yes," perform a hazard/risk analysis.

        If "no," coordinate with the Electrical Safety Program.

☑ ☐ **2.** Is a flash-hazard analysis required? (70E-2000, 2-1.3.3; 70E-2004, 130.3)

        3. Determine the shock protection boundary. (70E-2000 Table 2-1.3.4; 70E-2004 Table 130.2(B))
The restricted approach boundary distance is "avoid contact."

        4. Determine the flash protection boundary. (70E-2000, 2-1.3.3.2; 70E-2004, 130.3(A)(1))
The flash protection boundary distance is 4 ft, 0 in.

        5. Select the hazard/risk category for the task from Appendix G.
The hazard risk category is 1.

        6. Check notes on the table in Appendix G to determine the incident energy level for the specific task in the table.

☑ ☐ **7.** Is the fault energy level available at the location equal to or less than the table notes in Appendix G?

        Note: *The available fault current must be 25 kA or less, with 2-cycle clearing time to be applicable.*

        If "yes," the table is applicable.   If "no," further investigation is required.

        8. Check the legend in Appendix G for additional requirements. List requirements:

        ______________________________   ______________________________

        ______________________________   ______________________________

        ______________________________   ______________________________

        9. Determine the need for voltage-rated tools and voltage-rated gloves.

☑ ☐   Are voltage-rated tools required?

☑ ☐   Are voltage-rated gloves required?

        10. On 70E-2000 Table 3-3.9.2; 70E-2004 Table 130.7(C)(10) locate the hazard/risk category number. Select the PPE based on the table requirements. Read and comply with all applicable notes to ensure safety.

        PPE required:

Untreated cotton long pants (Note 4)	Hard hat
FR long-sleeved shirt	Safety glasses
FR pants (Note 4)	Leather gloves, as needed
Coveralls (Note 5)	Leather shoes, as needed

TASK ASSESSMENT CHECKLIST FOR COMMON TASKS
ON SYSTEMS RATED 600 VOLTS AND BELOW
Job Task E6: Voltage Testing at Utilization Equipment
Rated More Than 240 Volts up to 600 Volts

Date: ________________________ **Job Location:** ____

Yes **No**

☑ ☐   **1.** Is the equipment operating at more that 50 volts or is a shock hazard present?

         If "yes," perform a hazard/risk analysis.

         If "no," coordinate with the Electrical Safety Program.

☑ ☐   **2.** Is a flash-hazard analysis required? (70E-2000, 2-1.3.3; 70E-2004, 130.3)

         3. Determine the shock protection boundary. (70E-2000 Table 2-1.3.4; 70E-2004 Table 130.2(B))
         The restricted approach boundary distance is 1 ft, 0 in.

         4. Determine the flash protection boundary. (70E-2000, 2-1.3.3.2; 70E-2004, 130.3(A)(1))
         The flash protection boundary distance is 4 ft, 0 in.

         5. Select the hazard/risk category for the task from Appendix G.
         The hazard risk category is 2.*

         6. Check notes on the table in Appendix G to determine the incident energy level for the specific task in the table.

☑ ☐   **7.** Is the fault energy level available at the location equal to or less than the table notes in Appendix G?

         Note: *The available fault current must be 25 kA or less, with 2-cycle clearing time to be applicable.*

         If "yes," the table is applicable.   If "no," further investigation is required.

         8. Check the legend in Appendix G for additional requirements. List requirements:

         ______________________________   ______________________________

         ______________________________   ______________________________

         ______________________________   ______________________________

         9. Determine the need for voltage-rated tools and voltage-rated gloves.

☑ ☐   Are voltage-rated tools required?

☑ ☐   Are voltage-rated gloves required?

         10. On 70E-2000 Table 3-3.9.2; 70E-2004 Table 130.7(C)(10) locate the hazard/risk category number. Select the PPE based on the table requirements. Read and comply with all applicable notes to ensure safety.

         PPE required:

Untreated cotton t-shirt	Safety glasses or goggles
Untreated cotton long pants (Note 6)	Double layer switching hood
FR long-sleeved shirt (Note 9)	Hearing protection
FR long pants (Note 6)	Leather gloves
Coveralls (Note 7)	Leather shoes

TASK ASSESSMENT CHECKLIST FOR OTHER EQUIPMENT RATED AT 600 VOLTS
Job Task F1: Removing Bolted Covers (to Expose Bare, Energized Parts)

Date: _____________________   Job Location: _________________________________

Yes   No

☑ ☐   **1.** Is the equipment operating at more that 50 volts or is a shock hazard present?

      If "yes," perform a hazard/risk analysis.

      If "no," coordinate with the Electrical Safety Program.

☑ ☐   **2.** Is a flash-hazard analysis required? (70E-2000, 2-1.3.3; 70E-2004, 130.3)

      3. Determine the shock protection boundary. (70E-2000 Table 2-1.3.4; 70E-2004 Table 130.2(B))

      The restricted approach boundary distance is 1 ft, 0 in.

      4. Determine the flash protection boundary. (70E-2000, 2-1.3.3.2; 70E-2004, 130.3(A)(1))

      The flash protection boundary distance is 4 ft, 0 in.

      5. Select the hazard/risk category for the task from Appendix G.

      The hazard risk category is 2.*

      6. Check notes on the table in Appendix G to determine the incident energy level for the specific task in the table.

☑ ☐   **7.** Is the fault energy level available at the location equal to or less than the table notes in Appendix G?

      Note: *Must be 25 kA or less, with 2-cycle clearing time, to be applicable.*

      If "yes," the table is applicable.   If "no," further investigation is required.

      8. Check the legend in Appendix G for additional requirements. List requirements:

      _______________________________   _______________________________

      _______________________________   _______________________________

      _______________________________   _______________________________

      9. Determine the need for voltage-rated tools and voltage-rated gloves.

☐ ☑   Are voltage-rated tools required?

☐ ☑   Are voltage-rated gloves required?

      10. On 70E-2000 Table 3-3.9.2; 70E-2004 Table 130.7(C)(10) locate the hazard/risk category number. Select the PPE based on the table requirements. Read and comply with all applicable notes to ensure safety.

      PPE required:

Untreated cotton t-shirt	Safety glasses or goggles
Untreated cotton long pants (Note 6)	Double layer switching hood
FR long-sleeved shirt (Note 9)	Hearing protection
FR long pants (Note 6)	Leather gloves
Coveralls (Note 7)	Leather shoes
Hard hat	

TASK ASSESSMENT CHECKLIST FOR OTHER EQUIPMENT RATED AT 600 VOLTS

Job Task F2: Opening Hinged Covers (to Expose Bare, Energized Parts)

Date: ________________________   Job Location: ______________________________________

Yes   No

☑ ☐   **1.** Is the equipment operating at more that 50 volts or is a shock hazard present?

        If "yes," perform a hazard/risk analysis.

        If "no," coordinate with the Electrical Safety Program.

☑ ☐   **2.** Is a flash-hazard analysis required? (70E-2000, 2-1.3.3; 70E-2004, 130.3)

        3. Determine the shock protection boundary. (70E-2000 Table 2-1.3.4; 70E-2004 Table 130.2(B))

        The restricted approach boundary distance is 1 ft, 0 in.

        4. Determine the flash protection boundary. (70E-2000, 2-1.3.3.2; 70E-2004, 130.3(A)(1))

        The flash protection boundary distance is 4 ft, 0 in.

        5. Select the hazard/risk category for the task from Appendix G.

        The hazard risk category is 1.

        6. Check notes on the table in Appendix G to determine the incident energy level for the specific task in the table.

☑ ☐   **7.** Is the fault energy level available at the location equal to or less than the table notes in Appendix G?

        Note: Must be 25 kA or less, with 2-cycle clearing time, to be applicable.

        If "yes," the table is applicable.   If "no," further investigation is required.

        8. Check the legend in Appendix G for additional requirements. List requirements:

        __________________________   __________________________

        __________________________   __________________________

        __________________________   __________________________

        9. Determine the need for voltage-rated tools and voltage-rated gloves.

☐ ☑   Are voltage-rated tools required?

☐ ☑   Are voltage-rated gloves required?

        10. On 70E-2000 Table 3-3.9.2; 70E-2004 Table 130.7(C)(10) locate the hazard/risk category number. Select the PPE based on the table requirements. Read and comply with all applicable notes to ensure safety.

        PPE required:

Untreated cotton long pants (Note 4)	Hard hat
FR long-sleeved shirt	Safety glasses
FR pants (Note 4)	Leather gloves
Coveralls (Note 5)	Leather shoes

                                                    (p. 2 of 7)

TASK ASSESSMENT CHECKLIST FOR OTHER EQUIPMENT RATED AT 600 VOLTS

Job Task F3: Inserting or Removing Equipment (Such as a Revenue Meter)

Date: _______________________   **Job Location:** ____

Yes   No

☑ ☐   **1.** Is the equipment operating at more that 50 volts or is a shock hazard present?

     If "yes," perform a hazard/risk analysis.

     If "no," coordinate with the Electrical Safety Program.

☑ ☐   **2.** Is a flash-hazard analysis required? (70E-2000, 2-1.3.3; 70E-2004, 130.3)

     3. Determine the shock protection boundary. (70E-2000 Table 2-1.3.4; 70E-2004 Table 130.2(B))

     The restricted approach boundary distance is 1 ft, 0 in.

     4. Determine the flash protection boundary. (70E-2000, 2-1.3.3.2; 70E-2004, 130.3(A)(1))

     The flash protection boundary distance is 4 ft, 0 in.

     5. Select the hazard/risk category for the task from Appendix G.

     The hazard risk category is 2.*

     6. Check notes on the table in Appendix G to determine the incident energy level for the specific task in the table.

☑ ☐   **7.** Is the fault energy level available at the location equal to or less than the table notes in Appendix G?

     Note: Must be 25 kA or less, with 2-cycle clearing time, to be applicable.

     If "yes," the table is applicable.   If "no," further investigation is required.

     8. Check the legend in Appendix G for additional requirements. List requirements:

     _______________________________   _______________________________

     _______________________________   _______________________________

     _______________________________   _______________________________

     9. Determine the need for voltage-rated tools and voltage-rated gloves.

☐ ☑   Are voltage-rated tools required?

☑ ☐   Are voltage-rated gloves required?

     10. On 70E-2000 Table 3-3.9.2; 70E-2004 Table 130.7(C)(10) locate the hazard/risk category number. Select the PPE based on the table requirements. Read and comply with all applicable notes to ensure safety.

     PPE required:

Untreated cotton t-shirt	Safety glasses or goggles
Untreated cotton long pants (Note 6)	Double layer switching hood
FR long-sleeved shirt (Note 9)	Hearing protection
FR long pants (Note 6)	Leather gloves
Coveralls (Note 7)	Leather shoes
Hard hat	

       (p. 3 of 7)

TASK ASSESSMENT CHECKLIST FOR OTHER EQUIPMENT RATED AT 600 VOLTS
Job Task F4: Working on Energized Parts, Including Voltage Testing

Date: ________________________   Job Location: ________________________________

Yes   No

☑   ☐   **1.** Is the equipment operating at more that 50 volts or is a shock hazard present?

If "yes," perform a hazard/risk analysis.

If "no," coordinate with the Electrical Safety Program.

☑   ☐   **2.** Is a flash-hazard analysis required? (70E-2000, 2-1.3.3; 70E-2004, 130.3)

3. Determine the shock protection boundary. (70E-2000 Table 2-1.3.4; 70E-2004 Table 130.2(B))

The restricted approach boundary distance is 1 ft, 0 in.

4. Determine the flash protection boundary. (70E-2000, 2-1.3.3.2; 70E-2004, 130.3(A)(1))

The flash protection boundary distance is 4 ft, 0 in.

5. Select the hazard/risk category for the task from Appendix G.

The hazard risk category is 2.*

6. Check notes on the table in Appendix G to determine the incident energy level for the specific task in the table.

☑   ☐   **7.** Is the fault energy level available at the location equal to or less than the table notes in Appendix G?

Note: *Must be 25 kA or less, with 2-cycle clearing time, to be applicable.*

If "yes," the table is applicable.   If "no," further investigation is required.

8. Check the legend in Appendix G for additional requirements. List requirements:

_________________________________   _________________________________

_________________________________   _________________________________

_________________________________   _________________________________

9. Determine the need for voltage-rated tools and voltage-rated gloves.

☑   ☐   Are voltage-rated tools required?

☑   ☐   Are voltage-rated gloves required?

10. On 70E-2000 Table 3-3.9.2; 70E-2004 Table 130.7(C)(10) locate the hazard/risk category number. Select the PPE based on the table requirements. Read and comply with all applicable notes to ensure safety.

PPE required:

Untreated cotton t-shirt	Safety glasses or goggles
Untreated cotton long pants (Note 6)	Double layer switching hood
FR long-sleeved shirt (Note 9)	Hearing protection
FR long pants (Note 6)	Leather gloves, as needed
Coveralls (Note 7)	Leather shoes, as needed
Hard hat	

TASK ASSESSMENT CHECKLIST FOR OTHER EQUIPMENT RATED AT 600 VOLTS
Job Task F5: Installing or Removing Cable Trough or Tray Covers

Date: _________________________   **Job Location:** _________________________________

Yes   No

☑ ☐ **1.** Is the equipment operating at more that 50 volts or is a shock hazard present?

If "yes," perform a hazard/risk analysis.

If "no," coordinate with the Electrical Safety Program.

☑ ☐ **2.** Is a flash-hazard analysis required? (70E-2000, 2-1.3.3; 70E-2004, 130.3)

3. Determine the shock protection boundary. (70E-2000 Table 2-1.3.4; 70E-2004 Table 130.2(B))
The restricted approach boundary distance is 1 ft, 0 in.

4. Determine the flash protection boundary. (70E-2000, 2-1.3.3.2; 70E-2004, 130.3(A)(1))
The flash protection boundary distance is 4 ft, 0 in.

5. Select the hazard/risk category for the task from Appendix G.
The hazard risk category is 1.

6. Check notes on the table in Appendix G to determine the incident energy level for the specific task in the table.

☑ ☐ **7.** Is the fault energy level available at the location equal to or less than the table notes in Appendix G?

Note: Must be 25 kA or less, with 2-cycle clearing time, to be applicable.

If "yes," the table is applicable.   If "no," further investigation is required.

8. Check the legend in Appendix G for additional requirements. List requirements:

___________________________   ___________________________

___________________________   ___________________________

___________________________   ___________________________

9. Determine the need for voltage-rated tools and voltage-rated gloves.

☐ ☑ Are voltage-rated tools required?

☐ ☑ Are voltage-rated gloves required?

10. On 70E-2000 Table 3-3.9.2; 70E-2004 Table 130.7(C)(10) locate the hazard/risk category number. Select the PPE based on the table requirements. Read and comply with all applicable notes to ensure safety.

PPE required:

Untreated cotton long pants (Note 4)	Hard hat
FR long-sleeved shirt	Safety glasses
FR pants (Note 4)	Leather gloves, as needed
Coveralls (Note 5)	Leather shoes, as needed

    (p. 5 of 7)

TASK ASSESSMENT CHECKLIST FOR OTHER EQUIPMENT RATED AT 600 VOLTS

Job Task F6: Installing or Removing Miscellaneous Equipment Covers

Date: _________________________   Job Location: _______________________________

Yes   No

☑ ☐   **1.** Is the equipment operating at more that 50 volts or is a shock hazard present?

If "yes," perform a hazard/risk analysis.

If "no," coordinate with the Electrical Safety Program.

☑ ☐   **2.** Is a flash-hazard analysis required? (70E-2000, 2-1.3.3; 70E-2004, 130.3)

3. Determine the shock protection boundary. (70E-2000 Table 2-1.3.4; 70E-2004 Table 130.2(B))

The restricted approach boundary distance is 1 ft, 0 in.

4. Determine the flash protection boundary. (70E-2000, 2-1.3.3.2; 70E-2004, 130.3(A)(1))

The flash protection boundary distance is 4 ft, 0 in.

5. Select the hazard/risk category for the task from Appendix G.

The hazard risk category is 1.

6. Check notes on the table in Appendix G to determine the incident energy level for the specific task in the table.

☑ ☐   **7.** Is the fault energy level available at the location equal to or less than the table notes in Appendix G?

Note: *Must be 25 kA or less, with 2-cycle clearing time, to be applicable.*

If "yes," the table is applicable.   If "no," further investigation is required.

8. Check the legend in Appendix G for additional requirements. List requirements:

___________________________   ___________________________

___________________________   ___________________________

___________________________   ___________________________

9. Determine the need for voltage-rated tools and voltage-rated gloves.

☐ ☑   Are voltage-rated tools required?

☐ ☑   Are voltage-rated gloves required?

10. On 70E-2000 Table 3-3.9.2; 70E-2004 Table 130.7(C)(10) locate the hazard/risk category number. Select the PPE based on the table requirements. Read and comply with all applicable notes to ensure safety.

PPE required:

Untreated cotton long pants (Note 4)	Hard hat
FR long-sleeved shirt	Safety glasses
FR pants (Note 4)	Leather gloves, as needed
Coveralls (Note 5)	Leather shoes, as needed

TASK ASSESSMENT CHECKLIST FOR OTHER EQUIPMENT RATED AT 600 VOLTS
Job Task F7: Applying Safety Grounds after Voltage Testing

Date: _________________________   Job Location: _______________________________

Yes   No

☑ ☐ **1.** Is the equipment operating at more that 50 volts or is a shock hazard present?

If "yes," perform a hazard/risk analysis.

If "no," coordinate with the Electrical Safety Program.

☑ ☐ **2.** Is a flash-hazard analysis required? (70E-2000, 2-1.3.3; 70E-2004, 130.3)

3. Determine the shock protection boundary. (70E-2000 Table 2-1.3.4; 70E-2004 Table 130.2(B))

The restricted approach boundary distance is 1 ft, 0 in.

4. Determine the flash protection boundary. (70E-2000, 2-1.3.3.2; 70E-2004, 130.3(A)(1))

The flash protection boundary distance is 4 ft, 0 in.

5. Select the hazard/risk category for the task from Appendix G.

The hazard risk category is 2.*

6. Check notes on the table in Appendix G to determine the incident energy level for the specific task in the table.

☑ ☐ **7.** Is the fault energy level available at the location equal to or less than the table notes in Appendix G?

Note: *Must be 25 kA or less, with 2-cycle clearing time, to be applicable.*

If "yes," the table is applicable.   If "no," further investigation is required.

8. Check the legend in Appendix G for additional requirements. List requirements:

_____________________________   _____________________________

_____________________________   _____________________________

_____________________________   _____________________________

9. Determine the need for voltage-rated tools and voltage-rated gloves.

☐ ☑ Are voltage-rated tools required?

☑ ☐ Are voltage-rated gloves required?

10. On 70E-2000 Table 3-3.9.2; 70E-2004 Table 130.7(C)(10) locate the hazard/risk category number. Select the PPE based on the table requirements. Read and comply with all applicable notes to ensure safety.

PPE required:

Untreated cotton t-shirt	Safety glasses or goggles
Untreated cotton long pants (Note 6)	Double layer switching hood
FR long-sleeved shirt (Note 9)	Hearing protection
FR long pants (Note 6)	Leather gloves
Coveralls (Note 7)	Leather shoes
Hard hat	

TASK ASSESSMENT CHECKLIST FOR METAL-CLAD
SWITCHGEAR RATED AT 1 kV AND HIGHER

Job Task G1: Operating a Circuit Breaker or Fused Switch with the Enclosure Door Closed

Date: _________________________   **Job Location:** _______________________________

Yes   No

☑ ☐ **1.** Is the equipment operating at more that 50 volts or is a shock hazard present?

If "yes," perform a hazard/risk analysis.

If "no," coordinate with the Electrical Safety Program.

☑ ☐ **2.** Is a flash-hazard analysis required? (70E-2000, 2-1.3.3; 70E-2004, 130.3)

3. Determine the shock protection boundary. (70E-2000 Table 2-1.3.4; 70E-2004 Table 130.2(B))

4. Determine the flash protection boundary. (70E-2000, 2-1.3.3.2; 70E-2004, 130.3(A)(1))

The flash protection boundary distance is 20 ft, 0 in.

5. Select the hazard/risk category for the task from Appendix G.

The hazard risk category is 2.

6. Check notes on the table in Appendix G to determine the incident energy level for the specific task in the table.

☑ ☐ **7.** Is the fault energy level available at the location equal to or less than the table notes in Appendix G?

Note: *Not applicable unless the doors are louvered.*

If "yes," the table is applicable.   If "no," further investigation is required.

8. Check the legend in Appendix G for additional requirements. List requirements:

____________________________   ____________________________

____________________________   ____________________________

____________________________   ____________________________

9. Determine the need for voltage-rated tools and voltage-rated gloves.

☐ ☑ Are voltage-rated tools required?

☐ ☑ Are voltage-rated gloves required?

10. On 70E-2000 Table 3-3.9.2; 70E-2004 Table 130.7(C)(10) locate the hazard/risk category number. Select the PPE based on the table requirements. Read and comply with all applicable notes to ensure safety.

PPE required:

Untreated cotton t-shirt	Hard hat
Untreated cotton long pants (Note 6)	Safety glasses or goggles
FR long-sleeved shirt	Leather gloves
FR long pants (Note 6)	Leather shoes
Coveralls (Note 7)	

                              (p. 1 of 12)

TASK ASSESSMENT CHECKLIST FOR METAL-CLAD SWITCHGEAR RATED AT 1 kV AND HIGHER

Job Task G2: Reading a Panel Meter While Operating a Meter Switch

Date: ________________________   **Job Location:** ____

Yes   No

☑ ☐ **1.** Is the equipment operating at more that 50 volts or is a shock hazard present?

      If "yes," perform a hazard/risk analysis.

      If "no," coordinate with the Electrical Safety Program.

☑ ☐ **2.** Is a flash-hazard analysis required? (70E-2000, 2-1.3.3; 70E-2004, 130.3)

      3. Determine the shock protection boundary. (70E-2000 Table 2-1.3.4; 70E-2004 Table 130.2(B))

      4. Determine the flash protection boundary. (70E-2000, 2-1.3.3.2; 70E-2004, 130.3(A)(1))

      The flash protection boundary distance is 20 ft, 0 in.

      5. Select the hazard/risk category for the task from Appendix G.

      The hazard risk category is 0.

      6. Check notes on the table in Appendix G to determine the incident energy level for the specific task in the table.

☑ ☐ **7.** Is the fault energy level available at the location equal to or less than the table notes in Appendix G?

      Note: *Not applicable.*

      If "yes," the table is applicable.   If "no," further investigation is required.

      8. Check the legend in Appendix G for additional requirements. List requirements:

      ______________________________   ______________________________

      ______________________________   ______________________________

      ______________________________   ______________________________

      9. Determine the need for voltage-rated tools and voltage-rated gloves.

☐ ☑   Are voltage-rated tools required?

☐ ☑   Are voltage-rated gloves required?

      10. On 70E-2000 Table 3-3.9.2; 70E-2004 Table 130.7(C)(10) locate the hazard/risk category number. Select the PPE based on the table requirements. Read and comply with all applicable notes to ensure safety.

      PPE required:

      Untreated cotton long-sleeved shirt

      Untreated cotton long pants

      Safety glasses

**TASK ASSESSMENT CHECKLIST FOR METAL-CLAD
SWITCHGEAR RATED AT 1 kV AND HIGHER**

**Job Task G3: Operating a Circuit Breaker, Fused Switch,
or Starter with the Enclosure Door Open**

Date: ________________________   **Job Location:** ____

Yes *No*

☑ ☐ **1.** Is the equipment operating at more that 50 volts or is a shock hazard present?

        If "yes," perform a hazard/risk analysis.

        If "no," coordinate with the Electrical Safety Program.

☑ ☐ **2.** Is a flash-hazard analysis required? (70E-2000, 2-1.3.3; 70E-2004, 130.3)

        3. Determine the shock protection boundary. (70E-2000 Table 2-1.3.4; 70E-2004 Table 130.2(B))

        4. Determine the flash protection boundary. (70E-2000, 2-1.3.3.2; 70E-2004, 130.3(A)(1))

        The flash protection boundary distance is 20 ft, 0 in.

        5. Select the hazard/risk category for the task from Appendix G.

        The hazard risk category is 4.

        6. Check notes on the table in Appendix G to determine the incident energy level for the specific task in the table.

☑ ☐ **7.** Is the fault energy level available at the location equal to or less than the table notes in Appendix G?

        Note: *Not applicable.*

        If "yes," the table is applicable. If "no," further investigation is required.

        8. Check the legend in Appendix G for additional requirements. List requirements:

        ____________________________   ____________________________

        ____________________________   ____________________________

        ____________________________   ____________________________

        9. Determine the need for voltage-rated tools and voltage-rated gloves.

☐ ☑   Are voltage-rated tools required?

☐ ☑   Are voltage-rated gloves required?

        10. On 70E-2000 Table 3-3.9.2; 70E-2004 Table 130.7(C)(10) locate the hazard/risk category number. Select the PPE based on the table requirements. Read and comply with all applicable notes to ensure safety.

        PPE required:

T-shirt	Hard hat
Untreated natural fiber pants	FR hard hat liner
FR long-sleeved shirt	Safety glasses or goggles
FR long pants	Double layer switching hood
or FR coveralls in lieu of FR long-sleeved shirt and long pants	Hearing protection
	Leather gloves
Flash-suit jacket	Leather work shoes
Flash-suit long pants	

TASK ASSESSMENT CHECKLIST FOR METAL-CLAD SWITCHGEAR RATED AT 1 kV AND HIGHER

Job Task G4: Working on Energized Parts, Including Voltage Testing

Date: _______________________ **Job Location:** _________________________________

Yes   No

☑   ☐   **1.** Is the equipment operating at more that 50 volts or is a shock hazard present?

If "yes," perform a hazard/risk analysis.

If "no," coordinate with the Electrical Safety Program.7

☑   ☐   **2.** Is a flash-hazard analysis required? (70E-2000, 2-1.3.3; 70E-2004, 130.3)

3. Determine the shock protection boundary. (70E-2000 Table 2-1.3.4; 70E-2004 Table 130.2(B))

4. Determine the flash protection boundary. (70E-2000, 2-1.3.3.2; 70E-2004, 130.3(A)(1))

The flash protection boundary distance is 20 ft, 0 in.

5. Select the hazard/risk category for the task from Appendix G.

The hazard risk category is 4.

6. Check notes on the table in Appendix G to determine the incident energy level for the specific task in the table.

☑   ☐   **7.** Is the fault energy level available at the location equal to or less than the table notes in Appendix G?

Note: *Not applicable.*

If "yes," the table is applicable.   If "no," further investigation is required.

8. Check the legend in Appendix G for additional requirements. List requirements:

_________________________________   _________________________________

_________________________________   _________________________________

_________________________________   _________________________________

9. Determine the need for voltage-rated tools and voltage-rated gloves.

☑   ☐   Are voltage-rated tools required?

☑   ☐   Are voltage-rated gloves required?

10. On 70E-2000 Table 3-3.9.2; 70E-2004 Table 130.7(C)(10) locate the hazard/risk category number. Select the PPE based on the table requirements. Read and comply with all applicable notes to ensure safety.

PPE required:

T-shirt	Hard hat
Untreated natural fiber pants	FR hard hat liner
FR long-sleeved shirt	Safety glasses or goggles
FR long pants	Double layer switching hood
or FR coveralls in lieu of FR long-sleeved shirt and long pants	Hearing protection
	Leather gloves
Flash-suit jacket	Leather work shoes
Flash-suit long pants	

TASK ASSESSMENT CHECKLIST FOR METAL-CLAD
SWITCHGEAR RATED AT 1 kV AND HIGHER

Job Task G5: Working on Control Circuits with
Exposed Parts Energized at 120 Volts or Below

Date: _______________________   **Job Location:** ____

Yes   No

☑   ☐   **1.** Is the equipment operating at more that 50 volts or is a shock hazard present?

If "yes," perform a hazard/risk analysis.

If "no," coordinate with the Electrical Safety Program.

☑   ☐   **2.** Is a flash-hazard analysis required? (70E-2000, 2-1.3.3; 70E-2004, 130.3)

3. Determine the shock protection boundary. (70E-2000 Table 2-1.3.4; 70E-2004 Table 130.2(B))

4. Determine the flash protection boundary. (70E-2000, 2-1.3.3.2; 70E-2004, 130.3(A)(1))

The flash protection boundary distance is 20 ft, 0 in.

5. Select the hazard/risk category for the task from Appendix G.

The hazard risk category is 2.

6. Check notes on the table in Appendix G to determine the incident energy level for the specific task in the table.

☑   ☐   **7.** Is the fault energy level available at the location equal to or less than the table notes in Appendix G?

Note: *Not applicable*

If "yes," the table is applicable.   If "no," further investigation is required.

8. Check the legend in Appendix G for additional requirements. List requirements:

___________________________   ___________________________

___________________________   ___________________________

___________________________   ___________________________

9. Determine the need for voltage-rated tools and voltage-rated gloves.

☐   ☑   Are voltage-rated tools required?

☐   ☑   Are voltage-rated gloves required?

10. On 70E-2000 Table 3-3.9.2; 70E-2004 Table 130.7(C)(10) locate the hazard/risk category number. Select the PPE based on the table requirements. Read and comply with all applicable notes to ensure safety.

PPE required:

Untreated cotton t-shirt	Hard hat
Untreated cotton long pants (Note 6)	Safety glasses or goggles
FR long-sleeved shirt	Leather gloves
FR long pants (Note 6)	Leather work shoes
Coveralls (Note 7)	

                                                                 (p. 5 of 12)

**TASK ASSESSMENT CHECKLIST FOR METAL-CLAD
SWITCHGEAR RATED AT 1 kV AND HIGHER**

**Job Task G6: Working on Control Circuits with
Exposed Parts Energized at More Than 120 Volts**

Date: _______________________   **Job Location:** _________________________________

Yes	No	

Yes   No

☑ ☐   **1.** Is the equipment operating at more that 50 volts or is a shock hazard present?

   If "yes," perform a hazard/risk analysis.

   If "no," coordinate with the Electrical Safety Program.

☑ ☐   **2.** Is a flash-hazard analysis required? (70E-2000, 2-1.3.3; 70E-2004, 130.3)

   3. Determine the shock protection boundary. (70E-2000 Table 2-1.3.4; 70E-2004 Table 130.2(B))

   4. Determine the flash protection boundary. (70E-2000, 2-1.3.3.2; 70E-2004, 130.3(A)(1))

   The flash protection boundary distance is 20 ft, 0 in.

   5. Select the hazard/risk category for the task from Appendix G.

   The hazard risk category is 4.

   6. Check notes on the table in Appendix G to determine the incident energy level for the specific task in the table.

☑ ☐   **7.** Is the fault energy level available at the location equal to or less than the table notes in Appendix G?

   Note: *Not applicable.*

   If "yes," the table is applicable.   If "no," further investigation is required.

   8. Check the legend in Appendix G for additional requirements. List requirements:

   _______________________________   _______________________________

   _______________________________   _______________________________

   _______________________________   _______________________________

   9. Determine the need for voltage-rated tools and voltage-rated gloves.

☑ ☐   Are voltage-rated tools required?

☑ ☐   Are voltage-rated gloves required?

   10. On 70E-2000 Table 3-3.9.2; 70E-2004 Table 130.7(C)(10) locate the hazard/risk category number. Select the PPE based on the table requirements. Read and comply with all applicable notes to ensure safety.

   PPE required:

T-shirt	Hard hat
Untreated natural fiber pants	FR hard hat liner
FR long-sleeved shirt	Safety glasses or goggles
FR long pants	Double layer switching hood
or FR coveralls in lieu of FR long-sleeved shirt and long pants	Hearing protection
	Leather gloves
Flash-suit jacket	Leather work shoes
Flash-suit long pants	

TASK ASSESSMENT CHECKLIST FOR METAL-CLAD
SWITCHGEAR RATED AT 1 kV AND HIGHER
Job Task G7: Inserting or Removing (Racking) Circuit
Breakers from Cubicles with the Doors Closed

Date: _________________________   **Job Location:** ____

Yes   No

[✓] [ ]   **1.** Is the equipment operating at more that 50 volts or is a shock hazard present?

If "yes," perform a hazard/risk analysis.

If "no," coordinate with the Electrical Safety Program.

[✓] [ ]   **2.** Is a flash-hazard analysis required? (70E-2000, 2-1.3.3; 70E-2004, 130.3)

3. Determine the shock protection boundary. (70E-2000 Table 2-1.3.4; 70E-2004 Table 130.2(B))

4. Determine the flash protection boundary. (70E-2000, 2-1.3.3.2; 70E-2004, 130.3(A)(1))

The flash protection boundary distance is 20 ft, 0 in.

5. Select the hazard/risk category for the task from Appendix G.

The hazard risk category is 2.

6. Check notes on the table in Appendix G to determine the incident energy level for the specific task in the table.

[✓] [ ]   **7.** Is the fault energy levelv available at the location equal to or less than the table notes in Appendix G?

Note: *Not applicable.*

If "yes," the table is applicable.   If "no," further investigation is required.

8. Check the legend in Appendix G for additional requirements. List requirements:

___________________________   ___________________________

___________________________   ___________________________

____

9. Determine the need for voltage-rated tools and voltage-rated gloves.

[ ] [✓]   Are voltage-rated tools required?

[ ] [✓]   Are voltage-rated gloves required?

10. On 70E-2000 Table 3-3.9.2; 70E-2004 Table 130.7(C)(10) locate the hazard/risk category number. Select the PPE based on the table requirements. Read and comply with all applicable notes to ensure safety.

PPE required:

T-shirt	Hard hat
Untreated natural fiber pants	Safety glasses or goggles
FR long-sleeved shirt	Leather gloves
FR long pants (Note 6)	Leather work shoes
or FR coveralls in lieu of FR long-sleeved shirt and long pants	

                                      (p. 7 of 12)

TASK ASSESSMENT CHECKLIST FOR METAL-CLAD SWITCHGEAR RATED AT 1 kV AND HIGHER

Job Task G8: Inserting or Removing (Racking) Circuit Breakers from Cubicles with the Doors Open

Date: _________________________   **Job Location:** _______________________________________

Yes   *No*

☑   ☐   **1.** Is the equipment operating at more that 50 volts or is a shock hazard present?

            If "yes," perform a hazard/risk analysis.

            If "no," coordinate with the Electrical Safety Program.

☑   ☐   **2.** Is a flash-hazard analysis required? (70E-2000, 2-1.3.3; 70E-2004, 130.3)

            3. Determine the shock protection boundary. (70E-2000 Table 2-1.3.4; 70E-2004 Table 130.2(B))

            4. Determine the flash protection boundary. (70E-2000, 2-1.3.3.2; 70E-2004, 130.3(A)(1))

            The flash protection boundary distance is 20 ft, 0 in.

            5. Select the hazard/risk category for the task from Appendix G.

            The hazard risk category is 4.

            6. Check notes on the table in Appendix G to determine the incident energy level for the specific task in the table.

☑   ☐   **7.** Is the fault energy level available at the location equal to or less than the table notes in Appendix G?

            Note: *Not applicable.*

            If "yes," the table is applicable.   If "no," further investigation is required.

            8. Check the legend in Appendix G for additional requirements. List requirements:

            ______________________________   ______________________________

            ______________________________   ______________________________

            ______________________________   ______________________________

            9. Determine the need for voltage-rated tools and voltage-rated gloves.

☐   ☑   Are voltage-rated tools required?

☐   ☑   Are voltage-rated gloves required?

            10. On 70E-2000 Table 3-3.9.2; 70E-2004 Table 130.7(C)(10) locate the hazard/risk category number. Select the PPE based on the table requirements. Read and comply with all applicable notes to ensure safety.

            PPE required:

T-shirt	Hard hat
Untreated natural fiber pants	FR hard hat liner
FR long-sleeved shirt	Safety glasses or goggles
FR long pants	Double layer switching hood
or FR coveralls in lieu of FR long-sleeved shirt and long pants	Hearing protection
	Leather gloves
Flash-suit jacket	Leather work shoes
Flash-suit long pants	

TASK ASSESSMENT CHECKLIST FOR METAL-CLAD SWITCHGEAR RATED AT 1 kV AND HIGHER

Job Task G9: Removing Bolted Covers (to Expose Bare, Energized Parts)

Date: _________________________   **Job Location:** ____

Yes *No*

☑ ☐   **1.** Is the equipment operating at more that 50 volts or is a shock hazard present?

  If "yes," perform a hazard/risk analysis.

  If "no," coordinate with the Electrical Safety Program.

☑ ☐   **2.** Is a flash-hazard analysis required? (70E-2000, 2-1.3.3; 70E-2004, 130.3)

  3. Determine the shock protection boundary. (70E-2000 Table 2-1.3.4; 70E-2004 Table 130.2(B))

  4. Determine the flash protection boundary. (70E-2000, 2-1.3.3.2; 70E-2004, 130.3(A)(1))

  The flash protection boundary distance is 20 ft, 0 in.

  5. Select the hazard/risk category for the task from Appendix G.

  The hazard risk category is 4.

  6. Check notes on the table in Appendix G to determine the incident energy level for the specific task in the table.

☑ ☐   **7.** Is the fault energy level available at the location equal to or less than the table notes in Appendix G?

  Note: *Not applicable.*

  If "yes," the table is applicable.   If "no," further investigation is required.

  8. Check the legend in Appendix G for additional requirements. List requirements:

  ________________________________   ________________________________

  ________________________________   ________________________________

  ________________________________   ________________________________

  9. Determine the need for voltage-rated tools and voltage-rated gloves.

☐ ☑   Are voltage-rated tools required?

☐ ☑   Are voltage-rated gloves required?

  10. On 70E-2000 Table 3-3.9.2; 70E-2004 Table 130.7(C)(10) locate the hazard/risk category number. Select the PPE based on the table requirements. Read and comply with all applicable notes to ensure safety.

  PPE required:

T-shirt	Hard hat
Untreated natural fiber pants	FR hard hat liner
FR long-sleeved shirt	Safety glasses or goggles
FR long pants	Double layer switching hood
or FR coveralls in lieu of FR long-sleeved shirt and long pants	Hearing protection
	Leather gloves
Flash-suit jacket	Leather work shoes
Flash-suit long pants	

TASK ASSESSMENT CHECKLIST FOR METAL-CLAD
SWITCHGEAR RATED AT 1 kV AND HIGHER
Job Task G10: Opening Hinged Covers (to Expose Energized Parts)

Date: _________________________   **Job Location:** _______________________________________

Yes   No

☑   ☐   **1.** Is the equipment operating at more that 50 volts or is a shock hazard present?

If "yes," perform a hazard/risk analysis.

If "no," coordinate with the Electrical Safety Program.

☑   ☐   **2.** Is a flash-hazard analysis required? (70E-2000, 2-1.3.3; 70E-2004, 130.3)

3. Determine the shock protection boundary. (70E-2000 Table 2-1.3.4; 70E-2004 Table 130.2(B))

4. Determine the flash protection boundary. (70E-2000, 2-1.3.3.2; 70E-2004, 130.3(A)(1))

The flash protection boundary distance is 20 ft, 0 in.

5. Select the hazard/risk category for the task from Appendix G.

The hazard risk category is 3.

6. Check notes on the table in Appendix G to determine the incident energy level for the specific task in the table.

☑   ☐   **7.** Is the fault energy level available at the location equal to or less than the table notes in Appendix G?

Note: Not applicable.

If "yes," the table is applicable.   If "no," further investigation is required.

8. Check the legend in Appendix G for additional requirements. List requirements:

___________________________   ___________________________

___________________________   ___________________________

___________________________   ___________________________

9. Determine the need for voltage-rated tools and voltage-rated gloves.

☐   ☑   Are voltage-rated tools required?

☐   ☑   Are voltage-rated gloves required?

10. On 70E-2000 Table 3-3.9.2; 70E-2004 Table 130.7(C)(10) locate the hazard/risk category number. Select the PPE based on the table requirements. Read and comply with all applicable notes to ensure safety.

PPE required:

Untreated cotton t-shirt	Safety glasses or goggles
Untreated cotton long pants	Double layer switching hood
FR long-sleeved shirt (Note 9)	Hearing protection
FR long pants (Note 6)	Leather gloves
Hard hat	Leather work shoes
FR hard hat liner	

TASK ASSESSMENT CHECKLIST FOR METAL-CLAD SWITCHGEAR RATED AT 1 kV AND HIGHER

Job Task G11: Opening Voltage Transformer or Control Power Transformer Compartments

Date: ______________________   **Job Location:** ____

Yes   No

☑   ☐   **1.** Is the equipment operating at more that 50 volts or is a shock hazard present?

If "yes," perform a hazard/risk analysis.

If "no," coordinate with the Electrical Safety Program.

☑   ☐   **2.** Is a flash-hazard analysis required? (70E-2000, 2-1.3.3; 70E-2004, 130.3)

3. Determine the shock protection boundary. (70E-2000 Table 2-1.3.4; 70E-2004 Table 130.2(B))

4. Determine the flash protection boundary. (70E-2000, 2-1.3.3.2; 70E-2004, 130.3(A)(1))

The flash protection boundary distance is 20 ft, 0 in.

5. Select the hazard/risk category for the task from Appendix G.

The hazard risk category is 4.

6. Check notes on the table in Appendix G to determine the incident energy level for the specific task in the table.

☑   ☐   **7.** Is the fault energy level available at the location equal to or less than the table notes in Appendix G?

Note: *Not applicable.*

If "yes," the table is applicable.   If "no," further investigation is required.

8. Check the legend in Appendix G for additional requirements. List requirements:

_________________________________   _________________________________

_________________________________   _________________________________

_________________________________   _________________________________

9. Determine the need for voltage-rated tools and voltage-rated gloves.

☐   ☑   Are voltage-rated tools required?

☐   ☑   Are voltage-rated gloves required?

10. On 70E-2000 Table 3-3.9.2; 70E-2004 Table 130.7(C)(10) locate the hazard/risk category number. Select the PPE based on the table requirements. Read and comply with all applicable notes to ensure safety.

PPE required:

T-shirt	Hard hat
Untreated natural fiber pants	FR hard hat liner
FR long-sleeved shirt	Safety glasses or goggles
FR long pants	Double layer switching hood
or FR coveralls in lieu of FR long-sleeved shirt and long pants	Hearing protection
	Leather gloves
Flash-suit jacket	Leather work shoes
Flash-suit long pants	

TASK ASSESSMENT CHECKLIST FOR METAL-CLAD SWITCHGEAR RATED AT 1 kV AND HIGHER

Job Task G12: Applying Safety Grounds after Voltage Testing

Date: ________________________   **Job Location:** ____

Yes   No

☑ ☐ **1.** Is the equipment operating at more that 50 volts or is a shock hazard present?

   If "yes," perform a hazard/risk analysis.

   If "no," coordinate with the Electrical Safety Program.

☑ ☐ **2.** Is a flash-hazard analysis required? (70E-2000, 2-1.3.3; 70E-2004, 130.3)

3. Determine the shock protection boundary. (70E-2000 Table 2-1.3.4; 70E-2004 Table 130.2(B))

4. Determine the flash protection boundary. (70E-2000, 2-1.3.3.2; 70E-2004, 130.3(A)(1))

   The flash protection boundary distance is 20 ft, 0 in.

5. Select the hazard/risk category for the task from Appendix G.

   The hazard risk category is 4.

6. Check notes on the table in Appendix G to determine the incident energy level for the specific task in the table.

☑ ☐ **7.** Is the fault energy level available at the location equal to or less than the table notes in Appendix G?

   Note: *Not applicable.*

   If "yes," the table is applicable.   If "no," further investigation is required.

8. Check the legend in Appendix G for additional requirements. List requirements:

   ______________________________   ______________________________

   ______________________________   ______________________________

   ______________________________   ______________________________

9. Determine the need for voltage-rated tools and voltage-rated gloves.

☐ ☑   Are voltage-rated tools required?

☑ ☐   Are voltage-rated gloves required?

10. On 70E-2000 Table 3-3.9.2; 70E-2004 Table 130.7(C)(10) locate the hazard/risk category number. Select the PPE based on the table requirements. Read and comply with all applicable notes to ensure safety.

   PPE required:

T-shirt	Hard hat
Untreated natural fiber pants	FR hard hat liner
FR long-sleeved shirt	Safety glasses or goggles
FR long pants	Double layer switching hood
or FR coveralls in lieu of FR long-sleeved	Hearing protection
shirt and long pants	Leather gloves
Flash-suit jacket	Leather work shoes
Flash-suit long pants	

                                                                (p. 12 of 12)

TASK ASSESSMENT CHECKLIST FOR NEMA E2
MOTOR STARTER RATED FROM 2.3 kV TO 7.2 kV
Job Task H1: Reading a Panel Meter While Operating a Meter Switch

Date: ________________________   **Job Location:** ____

Yes	No	
☑	☐	**1.** Is the equipment operating at more that 50 volts or is a shock hazard present?

If "yes," perform a hazard/risk analysis.

If "no," coordinate with the Electrical Safety Program.

Yes	No	
☑	☐	**2.** Is a flash-hazard analysis required? (70E-2000, 2-1.3.3; 70E-2004, 130.3)

3. Determine the shock protection boundary. (70E-2000 Table 2-1.3.4; 70E-2004 Table 130.2(B))

4. Determine the flash protection boundary. (70E-2000, 2-1.3.3.2; 70E-2004, 130.3(A)(1))

The flash protection boundary distance is 20 ft, 0 in.

5. Select the hazard/risk category for the task from Appendix G.

The hazard risk category is 0.

6. Check notes on the table in Appendix G to determine the incident energy level for the specific task in the table.

Yes	No	
☑	☐	**7.** Is the fault energy level available at the location equal to or less than the table notes in Appendix G?

Note: *Not applicable.*

If "yes," the table is applicable.   If "no," further investigation is required.

8. Check the legend in Appendix G for additional requirements. List requirements:

________________________________   ________________________________

________________________________   ________________________________

________________________________   ________________________________

9. Determine the need for voltage-rated tools and voltage-rated gloves.

Yes	No	
☐	☑	Are voltage-rated tools required?
☐	☑	Are voltage-rated gloves required?

10. On 70E-2000 Table 3-3.9.2; 70E-2004 Table 130.7(C)(10) locate the hazard/risk category number. Select the PPE based on the table requirements. Read and comply with all applicable notes to ensure safety.

PPE required:

Untreated cotton long-sleeved shirt

Untreated cotton long pants

Safety glasses

TASK ASSESSMENT CHECKLIST FOR NEMA E2
MOTOR STARTER RATED FROM 2.3 kV TO 7.2 kV
Job Task H2: Operating a Contactor with the Enclosure Door Closed

Date: ________________________   **Job Location:** ____

Yes *No*

☑ ☐ **1.** Is the equipment operating at more that 50 volts or is a shock hazard present?

If "yes," perform a hazard/risk analysis.

If "no," coordinate with the Electrical Safety Program.

☑ ☐ **2.** Is a flash-hazard analysis required? (70E-2000, 2-1.3.3; 70E-2004, 130.3)

3. Determine the shock protection boundary. (70E-2000 Table 2-1.3.4; 70E-2004 Table 130.2(B))

4. Determine the flash protection boundary. (70E-2000, 2-1.3.3.2; 70E-2004, 130.3(A)(1))

The flash protection boundary distance is 20 ft, 0 in.

5. Select the hazard/risk category for the task from Appendix G.

The hazard risk category is 0.

6. Check notes on the table in Appendix G to determine the incident energy level for the specific task in the table.

☑ ☐ **7.** Is the fault energy level available at the location equal to or less than the table notes in Appendix G?

Note: *Not applicable.*

If "yes," the table is applicable.   If "no," further investigation is required.

8. Check the legend in Appendix G for additional requirements. List requirements:

________________________   ________________________

________________________   ________________________

________________________   ________________________

9. Determine the need for voltage-rated tools and voltage-rated gloves.

☐ ☑ Are voltage-rated tools required?

☐ ☑ Are voltage-rated gloves required?

10. On 70E-2000 Table 3-3.9.2; 70E-2004 Table 130.7(C)(10) locate the hazard/risk category number. Select the PPE based on the table requirements. Read and comply with all applicable notes to ensure safety.

PPE required:

Untreated cotton long-sleeved shirt

Untreated cotton long pants

Safety glasses

TASK ASSESSMENT CHECKLIST FOR NEMA E2
MOTOR STARTER RATED FROM 2.3 kV TO 7.2 kV

Job Task H3: Operating a Contactor with the Enclosure Door Open

Date: _________________________   Job Location: _________________________________

Yes *No*

☑ ☐ **1.** Is the equipment operating at more that 50 volts or is a shock hazard present?

If "yes," perform a hazard/risk analysis.

If "no," coordinate with the Electrical Safety Program.

☑ ☐ **2.** Is a flash-hazard analysis required? (70E-2000, 2-1.3.3; 70E-2004, 130.3)

3. Determine the shock protection boundary. (70E-2000 Table 2-1.3.4; 70E-2004 Table 130.2(B))

4. Determine the flash protection boundary. (70E-2000, 2-1.3.3.2; 70E-2004, 130.3(A)(1))

The flash protection boundary distance is 20 ft, 0 in.

5. Select the hazard/risk category for the task from Appendix G.

The hazard risk category is 2.*

6. Check notes on the table in Appendix G to determine the incident energy level for the specific task in the table.

☑ ☐ **7.** Is the fault energy level available at the location equal to or less than the table notes in Appendix G?

Note: *Not applicable.*

If "yes," the table is applicable.   If "no," further investigation is required.

8. Check the legend in Appendix G for additional requirements. List requirements:

_______________________________   _______________________________

_______________________________   _______________________________

_______________________________   _______________________________

9. Determine the need for voltage-rated tools and voltage-rated gloves.

☐ ☑ Are voltage-rated tools required?

☐ ☑ Are voltage-rated gloves required?

10. On 70E-2000 Table 3-3.9.2; 70E-2004 Table 130.7(C)(10) locate the hazard/risk category number. Select the PPE based on the table requirements. Read and comply with all applicable notes to ensure safety.

PPE required:

Untreated cotton t-shirt	Safety glasses or goggles
Untreated cotton long pants (Note 6)	Double layer switching hood
FR long-sleeved shirt (Note 9)	Hearing protection
FR long pants (Note 6)	Leather gloves
Coveralls (Note 7)	Leather shoes
Hard hat	

TASK ASSESSMENT CHECKLIST FOR NEMA E2
MOTOR STARTER RATED FROM 2.3 kV TO 7.2 kV

Job Task H4: Working on Energized Parts, Including Voltage Testing

Date: _________________________   Job Location: ____

Yes **No**

☑ ☐   **1.** Is the equipment operating at more that 50 volts or is a shock hazard present?

         If "yes," perform a hazard/risk analysis.

         If "no," coordinate with the Electrical Safety Program.

☑ ☐   **2.** Is a flash-hazard analysis required? (70E-2000, 2-1.3.3; 70E-2004, 130.3)

         3. Determine the shock protection boundary. (70E-2000 Table 2-1.3.4; 70E-2004 Table 130.2(B))

         4. Determine the flash protection boundary. (70E-2000, 2-1.3.3.2; 70E-2004, 130.3(A)(1))

         The flash protection boundary distance is 20 ft, 0 in.

         5. Select the hazard/risk category for the task from Appendix G.

         The hazard risk category is 3.

         6. Check notes on the table in Appendix G to determine the incident energy level for the specific task in the table.

☑ ☐   **7.** Is the fault energy level available at the location equal to or less than the table notes in Appendix G?

         Note: *Not applicable.*

         If "yes," the table is applicable.   If "no," further investigation is required.

         8. Check the legend in Appendix G for additional requirements. List requirements:

         _________________________________   _________________________________

         _________________________________   _________________________________

         _________________________________   _________________________________

         9. Determine the need for voltage-rated tools and voltage-rated gloves.

☑ ☐   Are voltage-rated tools required?

☑ ☐   Are voltage-rated gloves required?

       10. On 70E-2000 Table 3-3.9.2; 70E-2004 Table 130.7(C)(10) locate the hazard/risk category number. Select the PPE based on the table requirements. Read and comply with all applicable notes to ensure safety.

         PPE required:

Untreated cotton t-shirt	Safety glasses or goggles
Untreated cotton long pants	Double layer switching hood
FR long-sleeved shirt (Note 9)	Hearing protection
FR long pants (Note 9)	Leather gloves
Hard hat	Leather shoes
FR hard hat liner	

TASK ASSESSMENT CHECKLIST FOR NEMA E2
MOTOR STARTER RATED FROM 2.3 kV TO 7.2 kV

Job Task H5: Working on Control Circuits with Exposed Parts Energized at 120 Volts or Below

Date: ________________________   **Job Location:** ____

Yes **No**

☑ ☐ **1.** Is the equipment operating at more that 50 volts or is a shock hazard present?

   If "yes," perform a hazard/risk analysis.

   If "no," coordinate with the Electrical Safety Program.

☑ ☐ **2.** Is a flash-hazard analysis required? (70E-2000, 2-1.3.3; 70E-2004, 130.3)

   3. Determine the shock protection boundary. (70E-2000 Table 2-1.3.4; 70E-2004 Table 130.2(B))

   4. Determine the flash protection boundary. (70E-2000, 2-1.3.3.2; 70E-2004, 130.3(A)(1))

   The flash protection boundary distance is 20 ft, 0 in.

   5. Select the hazard/risk category for the task from Appendix G.

   The hazard risk category is 0.

   6. Check notes on the table in Appendix G to determine the incident energy level for the specific task in the table.

☑ ☐ **7.** Is the fault energy level available at the location equal to or less than the table notes in Appendix G?

   Note: *Not applicable.*

   If "yes," the table is applicable.   If "no," further investigation is required.

   8. Check the legend in Appendix G for additional requirements. List requirements:

   _________________________________   _________________________________

   _________________________________   _________________________________

   _________________________________   _________________________________

   9. Determine the need for voltage-rated tools and voltage-rated gloves.

☑ ☐   Are voltage-rated tools required?

☑ ☐   Are voltage-rated gloves required?

   10. On 70E-2000 Table 3-3.9.2; 70E-2004 Table 130.7(C)(10) locate the hazard/risk category number. Select the PPE based on the table requirements. Read and comply with all applicable notes to ensure safety.

   PPE required:

   Untreated cotton long-sleeved shirt

   Untreated cotton long pants

   Safety glasses

                                    (p. 5 of 11)

TASK ASSESSMENT CHECKLIST FOR NEMA E2
MOTOR STARTER RATED FROM 2.3 kV TO 7.2 kV

Job Task H6: Working on Control Circuits with Exposed Parts Energized at More Than 120 Volts

Date: ______________________   **Job Location:** ____

Yes   **No**

☑   ☐   **1.** Is the equipment operating at more that 50 volts or is a shock hazard present?

           If "yes," perform a hazard/risk analysis.

           If "no," coordinate with the Electrical Safety Program.

☑   ☐   **2.** Is a flash-hazard analysis required? (70E-2000, 2-1.3.3; 70E-2004, 130.3)

           3. Determine the shock protection boundary. (70E-2000 Table 2-1.3.4; 70E-2004 Table 130.2(B))

           4. Determine the flash protection boundary. (70E-2000, 2-1.3.3.2; 70E-2004, 130.3(A)(1))

           The flash protection boundary distance is 20 ft, 0 in.

           5. Select the hazard/risk category for the task from Appendix G.

           The hazard risk category is 3.

           6. Check notes on the table in Appendix G to determine the incident energy level for the specific task in the table.

☑   ☐   **7.** Is the fault energy level available at the location equal to or less than the table notes in Appendix G?

           Note: *Not applicable.*

           If "yes," the table is applicable.   If "no," further investigation is required.

           8. Check the legend in Appendix G for additional requirements. List requirements:

           _________________________________   _________________________________

           _________________________________   _________________________________

           _________________________________   _________________________________

           9. Determine the need for voltage-rated tools and voltage-rated gloves.

☑   ☐   Are voltage-rated tools required?

☑   ☐   Are voltage-rated gloves required?

           10. On 70E-2000 Table 3-3.9.2; 70E-2004 Table 130.7(C)(10) locate the hazard/risk category number. Select the PPE based on the table requirements. Read and comply with all applicable notes to ensure safety.

           PPE required:

Untreated cotton t-shirt	Safety glasses or goggles
Untreated cotton long pants	Double layer switching hood
FR long-sleeved shirt (Note 9)	Hearing protection
FR long pants (Note 9)	Leather gloves
Hard hat	Leather shoes
FR hard hat liner	

TASK ASSESSMENT CHECKLIST FOR NEMA E2
MOTOR STARTER RATED FROM 2.3 kV TO 7.2 kV

Job Task H7: Inserting or Removing (Racking) Circuit Breakers
from Cubicles with the Doors Open

Date: _________________________   Job Location: ____

Yes   No

☑ ☐   **1.** Is the equipment operating at more that 50 volts or is a shock hazard present?

       If "yes," perform a hazard/risk analysis.

       If "no," coordinate with the Electrical Safety Program.

☑ ☐   **2.** Is a flash-hazard analysis required? (70E-2000, 2-1.3.3; 70E-2004, 130.3)

       3. Determine the shock protection boundary. (70E-2000 Table 2-1.3.4; 70E-2004 Table 130.2(B))

       4. Determine the flash protection boundary. (70E-2000, 2-1.3.3.2; 70E-2004, 130.3(A)(1))

       The flash protection boundary distance is 20 ft, 0 in.

       5. Select the hazard/risk category for the task from Appendix G.

       The hazard risk category is 3.

       6. Check notes on the table in Appendix G to determine the incident energy level for the specific task in the table.

☑ ☐   **7.** Is the fault energy level available at the location equal to or less than the table notes in Appendix G?

       Note: *Not applicable.*

       If "yes," the table is applicable.   If "no," further investigation is required.

       8. Check the legend in Appendix G for additional requirements. List requirements:

       ____________________________   ____________________________

       ____________________________   ____________________________

       ____

       9. Determine the need for voltage-rated tools and voltage-rated gloves.

☐ ☑   Are voltage-rated tools required?

☐ ☑   Are voltage-rated gloves required?

       10. On 70E-2000 Table 3-3.9.2; 70E-2004 Table 130.7(C)(10) locate the hazard/risk category number. Select the PPE based on the table requirements. Read and comply with all applicable notes to ensure safety.

       PPE required:

Untreated cotton t-shirt	Safety glasses or goggles
Untreated cotton long pants	Double layer switching hood
FR long-sleeved shirt (Note 9)	Hearing protection
FR long pants (Note 9)	Leather gloves
Hard hat	Leather shoes
FR hard hat liner	

                                                                 (p. 7 of 11)

TASK ASSESSMENT CHECKLIST FOR NEMA E2
MOTOR STARTER RATED FROM 2.3 kV TO 7.2 kV

Job Task H8: Inserting or Removing (Racking) Circuit Breakers
from Cubicles with the Doors Closed

Date: ________________________   **Job Location:** ________________________________

Yes No

☑ ☐ **1.** Is the equipment operating at more that 50 volts or is a shock hazard present?

  If "yes," perform a hazard/risk analysis.

  If "no," coordinate with the Electrical Safety Program.

☑ ☐ **2.** Is a flash-hazard analysis required? (70E-2000, 2-1.3.3; 70E-2004, 130.3)

  3. Determine the shock protection boundary. (70E-2000 Table 2-1.3.4; 70E-2004 Table 130.2(B))

  4. Determine the flash protection boundary. (70E-2000, 2-1.3.3.2; 70E-2004, 130.3(A)(1))

  The flash protection boundary distance is 20 ft, 0 in.

  5. Select the hazard/risk category for the task from Appendix G.

  The hazard risk category is 2.

  6. Check notes on the table in Appendix G to determine the incident energy level for the specific task in the table.

☑ ☐ **7.** Is the fault energy level available at the location equal to or less than the table notes in Appendix G?

  Note: *Not applicable.*

  If "yes," the table is applicable.   If "no," further investigation is required.

  8. Check the legend in Appendix G for additional requirements. List requirements:

  ________________________________   ________________________________

  ________________________________   ________________________________

  ________________________________   ________________________________

  9. Determine the need for voltage-rated tools and voltage-rated gloves.

☑ ☐ Are voltage-rated tools required?

☑ ☐ Are voltage-rated gloves required?

  10. On 70E-2000 Table 3-3.9.2; 70E-2004 Table 130.7(C)(10) locate the hazard/risk category number. Select the PPE based on the table requirements. Read and comply with all applicable notes to ensure safety.

  PPE required:

  Untreated cotton t-shirt   Hard hat

  Untreated cotton long pants (Note 6)   Safety glasses or goggles

  FR long-sleeved shirt   Leather gloves

  FR long pants (Note 6)   Leather shoes

TASK ASSESSMENT CHECKLIST FOR NEMA E2
MOTOR STARTER RATED FROM 2.3 kV TO 7.2 kV

Job Task H9: Removing Bolted Covers (to Expose Bare, Energized Parts)

Date: _________________________   **Job Location:** _________________________________

Yes *No*

☑ ☐   **1.** Is the equipment operating at more that 50 volts or is a shock hazard present?

       If "yes," perform a hazard/risk analysis.

       If "no," coordinate with the Electrical Safety Program.

☑ ☐   **2.** Is a flash-hazard analysis required? (70E-2000, 2-1.3.3; 70E-2004, 130.3)

       3. Determine the shock protection boundary. (70E-2000 Table 2-1.3.4; 70E-2004 Table 130.2(B))

       4. Determine the flash protection boundary. (70E-2000, 2-1.3.3.2; 70E-2004, 130.3(A)(1))

       The flash protection boundary distance is 20 ft, 0 in.

       5. Select the hazard/risk category for the task from Appendix G.

       The hazard risk category is 4.

       6. Check notes on the table in Appendix G to determine the incident energy level for the specific task in the table.

☑ ☐   **7.** Is the fault energy level available at the location equal to or less than the table notes in Appendix G?

       Note: *Not applicable.*

       If "yes," the table is applicable.   If "no," further investigation is required.

       8. Check the legend in Appendix G for additional requirements. List requirements:

       __________________________   __________________________

       __________________________   __________________________

       __________________________   __________________________

       9. Determine the need for voltage-rated tools and voltage-rated gloves.

☐ ☑   Are voltage-rated tools required?

☐ ☑   Are voltage-rated gloves required?

       10. On 70E-2000 Table 3-3.9.2; 70E-2004 Table 130.7(C)(10) locate the hazard/risk category number. Select the PPE based on the table requirements. Read and comply with all applicable notes to ensure safety.

       PPE required:

Untreated cotton t-shirt	Hard hat
Untreated natural fiber pants	FR hard hat liner
FR long-sleeved shirt	Safety glasses or goggles
FR long pants	Double layer switching hood
or FR coveralls in lieu of FR long-sleeved	Hearing protection
shirt and long pants	Leather gloves
Flash-suit jacket	Leather work shoes
Flash-suit long pants	

TASK ASSESSMENT CHECKLIST FOR NEMA E2
MOTOR STARTER RATED FROM 2.3 kV TO 7.2 kV

Job Task H10: Opening Hinged Covers (to Expose Bare, Energized Parts)

Date: ___________________________   **Job Location:** ____

Yes   *No*

☑   ☐   **1.** Is the equipment operating at more that 50 volts or is a shock hazard present?

If "yes," perform a hazard/risk analysis.

If "no," coordinate with the Electrical Safety Program.

☑   ☐   **2.** Is a flash-hazard analysis required? (70E-2000, 2-1.3.3; 70E-2004, 130.3)

3. Determine the shock protection boundary. (70E-2000 Table 2-1.3.4; 70E-2004 Table 130.2(B))

4. Determine the flash protection boundary. (70E-2000, 2-1.3.3.2; 70E-2004, 130.3(A)(1))

The flash protection boundary distance is 20 ft, 0 in.

5. Select the hazard/risk category for the task from Appendix G.

The hazard risk category is 3.

6. Check notes on the table in Appendix G to determine the incident energy level for the specific task in the table.

☑   ☐   **7.** Is the fault energy level available at the location equal to or less than the table notes in Appendix G?

Note: *Not applicable.*

If "yes," the table is applicable.   If "no," further investigation is required.

8. Check the legend in Appendix G for additional requirements. List requirements:

_____________________________   _____________________________

_____________________________   _____________________________

_____________________________   _____________________________

9. Determine the need for voltage-rated tools and voltage-rated gloves.

☐   ☑   Are voltage-rated tools required?

☐   ☑   Are voltage-rated gloves required?

10. On 70E-2000 Table 3-3.9.2; 70E-2004 Table 130.7(C)(10) locate the hazard/risk category number. Select the PPE based on the table requirements. Read and comply with all applicable notes to ensure safety.

PPE required:

Untreated cotton t-shirt	Safety glasses or goggles
Untreated cotton long pants	Double layer switching hood
FR long-sleeved shirt (Note 9)	Hearing protection
FR long pants (Note 9)	Leather gloves
Hard hat	Leather shoes
FR hard hat liner	

TASK ASSESSMENT CHECKLIST FOR NEMA E2
MOTOR STARTER RATED FROM 2.3 kV TO 7.2 kV

Job Task H11: Applying Safety Grounds after Voltage Testing

Date: _________________________ Job Location: _______________________________

Yes   No

☑ ☐ **1.** Is the equipment operating at more that 50 volts or is a shock hazard present?

If "yes," perform a hazard/risk analysis.

If "no," coordinate with the Electrical Safety Program.

☑ ☐ **2.** Is a flash-hazard analysis required? (70E-2000, 2-1.3.3; 70E-2004, 130.3)

3. Determine the shock protection boundary. (70E-2000 Table 2-1.3.4; 70E-2004 Table 130.2(B))

4. Determine the flash protection boundary. (70E-2000, 2-1.3.3.2; 70E-2004, 130.3(A)(1))

The flash protection boundary distance is 20 ft, 0 in.

5. Select the hazard/risk category for the task from Appendix G.

The hazard risk category is 3.

6. Check notes on the table in Appendix G to determine the incident energy level for the specific task in the table.

☑ ☐ **7.** Is the fault energy level available at the location equal to or less than the table notes in Appendix G?

Note: *Not applicable.*

If "yes," the table is applicable.   If "no," further investigation is required.

8. Check the legend in Appendix G for additional requirements. List requirements:

_____________________________ _____________________________

_____________________________ _____________________________

_____________________________ _____________________________

9. Determine the need for voltage-rated tools and voltage-rated gloves.

☐ ☑ Are voltage-rated tools required?

☑ ☐ Are voltage-rated gloves required?

10. On 70E-2000 Table 3-3.9.2; 70E-2004 Table 130.7(C)(10) locate the hazard/risk category number. Select the PPE based on the table requirements. Read and comply with all applicable notes to ensure safety.

PPE required:

Untreated cotton t-shirt	Safety glasses or goggles
Untreated cotton long pants	Double layer switching hood
FR long-sleeved shirt (Note 9)	Hearing protection
FR long pants (Note 9)	Leather gloves
Hard hat	Leather shoes

TASK ASSESSMENT CHECKLIST FOR OTHER EQUIPMENT RATED AT 1 kV AND MORE

Job Task I1: Operating a Switch with the Enclosure Door Closed

Date: ________________________   Job Location: ________________________________

Yes   No

☑   ☐   **1.** Is the equipment operating at more that 50 volts or is a shock hazard present?

If "yes," perform a hazard/risk analysis.

If "no," coordinate with the Electrical Safety Program.

☑   ☐   **2.** Is a flash-hazard analysis required? (70E-2000, 2-1.3.3; 70E-2004, 130.3)

3. Determine the shock protection boundary. (70E-2000 Table 2-1.3.4; 70E-2004 Table 130.2(B))

4. Determine the flash protection boundary. (70E-2000, 2-1.3.3.2; 70E-2004, 130.3(A)(1))

The flash protection boundary distance is 20 ft, 0 in.

5. Select the hazard/risk category for the task from Appendix G.

The hazard risk category is 2.

6. Check notes on the table in Appendix G to determine the incident energy level for the specific task in the table.

☑   ☐   **7.** Is the fault energy level available at the location equal to or less than the table notes in Appendix G?

Note: *Not applicable.*

If "yes," the table is applicable.   If "no," further investigation is required.

8. Check the legend in Appendix G for additional requirements. List requirements:

________________________________   ________________________________

________________________________   ________________________________

________________________________   ________________________________

9. Determine the need for voltage-rated tools and voltage-rated gloves.

☐   ☑   Are voltage-rated tools required?

☐   ☑   Are voltage-rated gloves required?

10. On 70E-2000 Table 3-3.9.2; 70E-2004 Table 130.7(C)(10) locate the hazard/risk category number. Select the PPE based on the table requirements. Read and comply with all applicable notes to ensure safety.

PPE required:

Untreated cotton t-shirt   Hard hat

Untreated cotton long pants (Note 6)   Safety glasses or goggles

FR long-sleeved shirt   Leather gloves

FR long pants (Note 6)   Leather shoes

Coveralls (Note 7)

TASK ASSESSMENT CHECKLIST FOR OTHER EQUIPMENT RATED AT 1 kV AND MORE
Job Task I2: Working on Energized Parts, Including Voltage Testing

Date: _______________________   **Job Location:** ____

Yes **No**

☑ ☐ **1.** Is the equipment operating at more that 50 volts or is a shock hazard present?

If "yes," perform a hazard/risk analysis.

If "no," coordinate with the Electrical Safety Program.

☑ ☐ **2.** Is a flash-hazard analysis required? (70E-2000, 2-1.3.3; 70E-2004, 130.3)

3. Determine the shock protection boundary. (70E-2000 Table 2-1.3.4; 70E-2004 Table 130.2(B))

4. Determine the flash protection boundary. (70E-2000, 2-1.3.3.2; 70E-2004, 130.3(A)(1))

The flash protection boundary distance is 20 ft, 0 in.

5. Select the hazard/risk category for the task from Appendix G.

The hazard risk category is 4.

6. Check notes on the table in Appendix G to determine the incident energy level for the specific task in the table.

☑ ☐ **7.** Is the fault energy level available at the location equal to or less than the table notes in Appendix G?

Note: *Not applicable.*

If "yes," the table is applicable.   If "no," further investigation is required.

8. Check the legend in Appendix G for additional requirements. List requirements:

___________________________   ___________________________________

___________________________   ___________________________________

___________________________   ___________________________________

9. Determine the need for voltage-rated tools and voltage-rated gloves.

☑ ☐ Are voltage-rated tools required?

☑ ☐ Are voltage-rated gloves required?

10. On 70E-2000 Table 3-3.9.2; 70E-2004 Table 130.7(C)(10) locate the hazard/risk category number. Select the PPE based on the table requirements. Read and comply with all applicable notes to ensure safety.

PPE required:

T-shirt	Hard hat
Untreated natural fiber pants	FR hard hat liner
FR long-sleeved shirt	Safety glasses or goggles
FR long pants	Double layer switching hood
or FR coveralls in lieu of FR long-sleeved shirt and long pants	Hearing protection
	Leather gloves
Flash-suit jacket	Leather work shoes
Flash-suit long pants	

                                        (p. 2 of 8)

TASK ASSESSMENT CHECKLIST FOR OTHER EQUIPMENT RATED AT 1 kV AND MORE
Job Task 13: Removing Bolted Covers (to Expose Bare, Energized Parts)

Date: _________________________   Job Location: ____

Yes No

☑ ☐ **1.** Is the equipment operating at more that 50 volts or is a shock hazard present?

If "yes," perform a hazard/risk analysis.

If "no," coordinate with the Electrical Safety Program.

☑ ☐ **2.** Is a flash-hazard analysis required? (70E-2000, 2-1.3.3; 70E-2004, 130.3)

3. Determine the shock protection boundary. (70E-2000 Table 2-1.3.4; 70E-2004 Table 130.2(B))

4. Determine the flash protection boundary. (70E-2000, 2-1.3.3.2; 70E-2004, 130.3(A)(1))

The flash protection boundary distance is 20 ft, 0 in.

5. Select the hazard/risk category for the task from Appendix G.

The hazard risk category is 4.

6. Check notes on the table in Appendix G to determine the incident energy level for the specific task in the table.

☑ ☐ **7.** Is the fault energy level available at the location equal to or less than the table notes in Appendix G?

Note: *Not applicable.*

If "yes," the table is applicable.   If "no," further investigation is required.

8. Check the legend in Appendix G for additional requirements. List requirements:

_________________________________   _________________________________

_________________________________   _________________________________

_________________________________   _________________________________

9. Determine the need for voltage-rated tools and voltage-rated gloves.

☐ ☑ Are voltage-rated tools required?

☐ ☑ Are voltage-rated gloves required?

10. On 70E-2000 Table 3-3.9.2; 70E-2004 Table 130.7(C)(10) locate the hazard/risk category number. Select the PPE based on the table requirements. Read and comply with all applicable notes to ensure safety.

PPE required:

T-shirt	Hard hat
Untreated natural fiber pants	FR hard hat liner
FR long-sleeved shirt	Safety glasses or goggles
FR long pants	Double layer switching hood
or FR coveralls in lieu of FR long-sleeved shirt and long pants	Hearing protection
	Leather gloves
Flash-suit jacket	Leather work shoes
Flash-suit long pants	

TASK ASSESSMENT CHECKLIST FOR OTHER EQUIPMENT RATED AT 1 kV AND MORE

Job Task I4: Opening Hinged Covers (to Expose Bare, Energized Parts)

Date: _________________________   **Job Location:** _____________________________________

Yes   *No*

☑   ☐   **1.** Is the equipment operating at more that 50 volts or is a shock hazard present?

If "yes," perform a hazard/risk analysis.

If "no," coordinate with the Electrical Safety Program.

☑   ☐   **2.** Is a flash-hazard analysis required? (70E-2000, 2-1.3.3; 70E-2004, 130.3)

3. Determine the shock protection boundary. (70E-2000 Table 2-1.3.4; 70E-2004 Table 130.2(B))

4. Determine the flash protection boundary. (70E-2000, 2-1.3.3.2; 70E-2004, 130.3(A)(1))

The flash protection boundary distance is 20 ft, 0 in.

5. Select the hazard/risk category for the task from Appendix G.

The hazard risk category is 3.

6. Check notes on the table in Appendix G to determine the incident energy level for the specific task in the table.

☑   ☐   **7.** Is the fault energy level available at the location equal to or less than the table notes in Appendix G?

Note: *Not applicable.*

If "yes," the table is applicable.   If "no," further investigation is required.

8. Check the legend in Appendix G for additional requirements. List requirements:

_________________________________   _________________________________

_________________________________   _________________________________

_________________________________   _________________________________

9. Determine the need for voltage-rated tools and voltage-rated gloves.

☐   ☑   Are voltage-rated tools required?

☐   ☑   Are voltage-rated gloves required?

10. On 70E-2000 Table 3-3.9.2; 70E-2004 Table 130.7(C)(10) locate the hazard/risk category number. Select the PPE based on the table requirements. Read and comply with all applicable notes to ensure safety.

PPE required:

Untreated cotton t-shirt	Safety glasses or goggles
Untreated cotton long pants	Double layer switching hood
FR long-sleeved shirt (Note 9)	Hearing protection
FR long pants (Note 9)	Leather gloves
Hard hat	Leather work shoes
FR hard hat liner	

TASK ASSESSMENT CHECKLIST FOR OTHER EQUIPMENT RATED AT 1 kV AND MORE

Job Task I5: Operating an Outdoor Disconnect Switch with a Hookstick

Date: _________________________   Job Location: ________________________________

Yes   No

☑ ☐   **1.** Is the equipment operating at more that 50 volts or is a shock hazard present?

   If "yes," perform a hazard/risk analysis.

   If "no," coordinate with the Electrical Safety Program.

☑ ☐   **2.** Is a flash-hazard analysis required? (70E-2000, 2-1.3.3; 70E-2004, 130.3)

   3. Determine the shock protection boundary. (70E-2000 Table 2-1.3.4; 70E-2004 Table 130.2(B))

   4. Determine the flash protection boundary. (70E-2000, 2-1.3.3.2; 70E-2004, 130.3(A)(1))

   The flash protection boundary distance is 20 ft, 0 in.

   5. Select the hazard/risk category for the task from Appendix G.

   The hazard risk category is 3.

   6. Check notes on the table in Appendix G to determine the incident energy level for the specific task in the table.

☑ ☐   **7.** Is the fault energy level available at the location equal to or less than the table notes in Appendix G?

   Note: *Not applicable.*

   If "yes," the table is applicable.   If "no," further investigation is required.

   8. Check the legend in Appendix G for additional requirements. List requirements:

   _____________________________________   _____________________________________

   _____________________________________   _____________________________________

   _____________________________________   _____________________________________

   9. Determine the need for voltage-rated tools and voltage-rated gloves.

☑ ☐   Are voltage-rated tools required?

☑ ☐   Are voltage-rated gloves required?

   10. On 70E-2000 Table 3-3.9.2; 70E-2004 Table 130.7(C)(10) locate the hazard/risk category number. Select the PPE based on the table requirements. Read and comply with all applicable notes to ensure safety.

   PPE required:

Untreated cotton t-shirt	Safety glasses or goggles
Untreated cotton long pants	Double layer switching hood
FR long-sleeved shirt (Note 9)	Hearing protection
FR long pants (Note 9)	Leather gloves
Hard hat	Leather work shoes
FR hard hat liner	

TASK ASSESSMENT CHECKLIST FOR OTHER EQUIPMENT RATED AT 1 kV AND MORE

Job Task I6: Operating an Outdoor Disconnect Switch (Gang Operated, from Grade)

Date: _________________________   **Job Location:** ____

Yes *No*

☑   ☐   **1.** Is the equipment operating at more that 50 volts or is a shock hazard present?

        If "yes," perform a hazard/risk analysis.

        If "no," coordinate with the Electrical Safety Program.

☑   ☐   **2.** Is a flash-hazard analysis required? (70E-2000, 2-1.3.3; 70E-2004, 130.3)

        3. Determine the shock protection boundary. (70E-2000 Table 2-1.3.4; 70E-2004 Table 130.2(B))

        4. Determine the flash protection boundary. (70E-2000, 2-1.3.3.2; 70E-2004, 130.3(A)(1))

        The flash protection boundary distance is 20 ft, 0 in.

        5. Select the hazard/risk category for the task from Appendix G.

        The hazard risk category is 2.

        6. Check notes on the table in Appendix G to determine the incident energy level for the specific task in the table.

☑   ☐   **7.** Is the fault energy level available at the location equal to or less than the table notes in Appendix G?

        Note: *Not applicable.*

        If "yes," the table is applicable.   If "no," further investigation is required.

        8. Check the legend in Appendix G for additional requirements. List requirements:

        _________________________________   _________________________________

        _________________________________   _________________________________

        _________________________________   _________________________________

        9. Determine the need for voltage-rated tools and voltage-rated gloves.

☐   ☑   Are voltage-rated tools required?

☐   ☑   Are voltage-rated gloves required?

       10. On 70E-2000 Table 3-3.9.2; 70E-2004 Table 130.7(C)(10) locate the hazard/risk category number. Select the PPE based on the table requirements. Read and comply with all applicable notes to ensure safety.

        PPE required:

Untreated cotton t-shirt	Hard hat
Untreated cotton long pants (Note 6)	Safety glasses or goggles
FR long-sleeved shirt (Note 9)	Leather gloves
FR long pants (Note 9)	Leather work shoes
FR Coveralls (Note 9)	

      (p. 6 of 8)

TASK ASSESSMENT CHECKLIST FOR OTHER EQUIPMENT RATED AT 1 kV AND MORE
Job Task I7: Examining Insulated Cable in a Manhole or Other Confined Space

Date: _________________________   **Job Location:** ____

Yes *No*

☑ ☐   **1.** Is the equipment operating at more that 50 volts or is a shock hazard present?

If "yes," perform a hazard/risk analysis.

If "no," coordinate with the Electrical Safety Program.

☑ ☐   **2.** Is a flash-hazard analysis required? (70E-2000, 2-1.3.3; 70E-2004, 130.3)

3. Determine the shock protection boundary. (70E-2000 Table 2-1.3.4; 70E-2004 Table 130.2(B))

4. Determine the flash protection boundary. (70E-2000, 2-1.3.3.2; 70E-2004, 130.3(A)(1))

The flash protection boundary distance is 20 ft, 0 in.

5. Select the hazard/risk category for the task from Appendix G.

The hazard risk category is 4.

6. Check notes on the table in Appendix G to determine the incident energy level for the specific task in the table.

☑ ☐   **7.** Is the fault energy level available at the location equal to or less than the table notes in Appendix G?

Note: *Not applicable.*

If "yes," the table is applicable.   If "no," further investigation is required.

8. Check the legend in Appendix G for additional requirements. List requirements:

_________________________________   _________________________________

_________________________________   _________________________________

_________________________________   _________________________________

9. Determine the need for voltage-rated tools and voltage-rated gloves.

☐ ☑   Are voltage-rated tools required?

☑ ☐   Are voltage-rated gloves required?

10. On 70E-2000 Table 3-3.9.2; 70E-2004 Table 130.7(C)(10) locate the hazard/risk category number. Select the PPE based on the table requirements. Read and comply with all applicable notes to ensure safety.

PPE required:

T-shirt	Hard hat
Untreated natural fiber pants	FR hard hat liner
FR long-sleeved shirt	Safety glasses or goggles
FR long pants	Double layer switching hood
or FR coveralls in lieu of FR long-sleeved shirt and long pants	Hearing protection
	Leather gloves
Flash-suit jacket	Leather work shoes
Flash-suit long pants	

TASK ASSESSMENT CHECKLIST FOR OTHER EQUIPMENT RATED AT 1 kV AND MORE

Job Task I8: Examining Insulated Cable in Open Air

Date: _______________________   Job Location: ____

Yes   No

☑   ☐   **1.** Is the equipment operating at more that 50 volts or is a shock hazard present?

If "yes," perform a hazard/risk analysis.

If "no," coordinate with the Electrical Safety Program.

☑   ☐   **2.** Is a flash-hazard analysis required? (70E-2000, 2-1.3.3; 70E-2004, 130.3)

3. Determine the shock protection boundary. (70E-2000 Table 2-1.3.4; 70E-2004 Table 130.2(B))

4. Determine the flash protection boundary. (70E-2000, 2-1.3.3.2; 70E-2004, 130.3(A)(1))

The flash protection boundary distance is 20 ft, 0 in.

5. Select the hazard/risk category for the task from Appendix G.

The hazard risk category is 2.

6. Check notes on the table in Appendix G to determine the incident energy level for the specific task in the table.

☑   ☐   **7.** Is the fault energy level available at the location equal to or less than the table notes in Appendix G?

Note: Not applicable.

If "yes," the table is applicable.   If "no," further investigation is required.

8. Check the legend in Appendix G for additional requirements. List requirements:

________________________   ________________________

________________________   ________________________

________________________   ________________________

9. Determine the need for voltage-rated tools and voltage-rated gloves.

☐   ☑   Are voltage-rated tools required?

☑   ☐   Are voltage-rated gloves required?

10. On 70E-2000 Table 3-3.9.2; 70E-2004 Table 130.7(C)(10) locate the hazard/risk category number. Select the PPE based on the table requirements. Read and comply with all applicable notes to ensure safety.

PPE required:

Untreated cotton t-shirt	Hard hat
Untreated cotton long pants (Note 6)	FR hard hat liner
FR long-sleeved shirt	Safety glasses or goggles
FR long pants (Note 6)	Leather gloves
Coveralls (Note 7)	Leather work shoes

                                                          (p. 8 of 8)

Appendix D

Job Briefing and Planning Checklist

This appendix contains a job briefing and planning checklist that can be used in the company's electrical safety program. As discussed in Chapters 3, 5, and 12, the checklist identifies points that the electrical safety manager or area supervisor and the worker should include in their discussion when the task is assigned and before the work is begun.

Use of the specific information in the job briefing and planning checklist will identify the hazards and constraints involved in the task and ensure that essential planning occurs.

JOB BRIEFING AND PLANNING CHECKLIST

Identify

- ☐ The hazards
- ☐ The voltage levels involved
- ☐ Skills required
- ☐ Any "foreign" (secondary source) voltage source
- ☐ Any unusual work conditions
- ☐ How many people are needed to do the job
- ☐ The shock protection boundaries
- ☐ The available incident energy
- ☐ Potential for arc flash (Conduct a flash-hazard analysis.)
- ☐ The flash protection boundaries

Ask

- ☐ Can the equipment be de-energized?
- ☐ Are backfeeds of the circuits to be worked on possible?
- ☐ Is a "standby person" required?

Check

- ☐ Job plans
- ☐ Single-line diagrams and vendor prints
- ☐ Status board
- ☐ That information on plant and vendor resources is up to date
- ☐ Safety procedures
- ☐ Vendor information
- ☐ That individuals are familiar with the facility

Know

- ☐ What the job is
- ☐ Who else needs to know—communicate!
- ☐ Who is in charge

Think

- ☐ About the unexpected event…What if?
- ☐ Lock—Tag—Test—Try
- ☐ Test for voltage—FIRST
- ☐ Use the right tools and equipment, including PPE
- ☐ Install and remove grounds
- ☐ Install barriers and barricades
- ☐ What else…?

Prepare for an emergency

- ☐ Is the standby person CPR trained?
- ☐ Is the required emergency equipment available? Where is it?
- ☐ Where is the nearest telephone?
- ☐ Where is the fire alarm?
- ☐ Is confined space rescue available?
- ☐ What is the exact work location?
- ☐ How is the equipment shut off in an emergency?
- ☐ Are emergency telephone numbers known?
- ☐ Where is the fire extinguisher?
- ☐ Are radio communications available?

    (p. 1 of 1)

Appendix E

Energized Electrical Work Permit

This appendix provides an energized electrical work permit that can be used in the company's electrical safety program to ensure that the hazards of working on or near exposed live parts receive adequate consideration. The permit, which relates to topics covered in Chapters 5, 6, 7, and 12, advises both workers and equipment owners, in writing, that work is going to be performed while the circuit remains energized. *An equipment owner might be more likely to think about the consequences of his or her decision if he or she must sign a form accepting responsibility for that decision.* Using this permit assures the worker that the increased costs associated with work on or near an exposed electrical conductor that is energized are justified. The use and very existence of this permit can also sometimes help management to understand that work performed on or near exposed energized parts is not worth the risk. He or she might, then, find a way to shut down the system and perform the work on de-energized equipment.

ENERGIZED ELECTRICAL WORK PERMIT
Part I—Work Request
(To be completed by the person requesting the permit)

1. Site: _________________________   Area: _________________________

2. Work order/project #: ____

3. Planned start date: _________________   Time: _________   Duration: _________

4. Description of the work to be done: _________________________

____

____

5. Work classification:

☐ Prohibited   ☐ Restricted

6. The following equipment was requested to be shut down: _________________________

____

____

☐ Until work is complete   ☐ Temporarily, while barriers are being placed

7. Requested by:

_________________________   _________________   _________________

(Signature)   (Title)   (Date)

Part II—Justification for Request
(To be completed by the electrically qualified persons doing the work)

1. Detailed job description procedure to be used in performing the above described work: _________________

____

____

2. Description of the safe work practices to be employed: _________________

____

3. Results of the shock hazard analysis: _________________

____

4. Determination of shock protection boundaries: _________________

____

5. Results of the flash hazard analysis: _________________

____

6. Determination of the flash protection boundary: _________________

____

                                        (p. 1 of 2)

7. Necessary personal protective equipment to safely perform the assigned task: _________________

____

8. Means employed to restrict the access of unqualified persons from the work area: _____________

____

9. Evidence of completion of a job briefing, including discussion of any job-specific task: ___________

____

10. Do you agree that the work described above can be done safely?

☐ Yes   ☐ No

_______________________________   _______________________

(Signature, Electrically Qualified Person)   (Date)

_______________________________   _______________________

(Signature, Electrically Qualified Person)   (Date)

Part III—Approval to Perform the Work While Electrically Energized

(To be completed by operations)

1. Reason for live work request: ____

____

____

2. The next available date for shutdown is: ____

3. I deny the request for shutdown and authorize the live work to be done.

_______________________________   _______________________

(Signature, Operations Manager)   (Date)

4. Live work on this equipment is:

☐ Approved   ☐ Not approved

_______________________________   _______________________

(Signature, Manufacturing Manager)   (Date)

_______________________________   _______________________

(Signature, Safety Manager)   (Date)

_______________________________   _______________________

(Signature, General Manager)   (Date)

_______________________________   _______________________

(Signature, Maintenance/Engineering Manager)   (Date)

_______________________________   _______________________

(Electrically Knowledgeable Person)   (Date)

Appendix F

Sample Lockout/ Tagout Procedure

This appendix provides a sample lockout/tagout procedure that could be used in a company's electrical safety program. The sample procedure is based on requirements that are found in NFPA 70E, *Standard for Electrical Safety Requirements for Employee Workplaces.* This appendix relates primarily to Chapter 9, Lockout/Tagout, but the information it contains can also be used in conjunction with Chapter 4, Procedures and Plans, and Chapter 5, Site Assessment.

OSHA regulations identify lockout/tagout requirements in several different sections. Basic OSHA lockout/tagout requirements are contained in 29 CFR 1910.147. Lockout and tagout are also addressed in 29 CFR 1910.333, 1910.269, and 1926 Subpart K. Other consensus standards also contain lockout and tagout requirements and concerns. While each of these standards is founded in reality, differences are significant. These differences often result in confusion and misunderstanding of the lockout/tagout process.

The sample procedure provided in this appendix is intended to address all the requirements from these regulations and standards from the perspective of exposure to electrical hazards. This procedure can be used for individual employee control or simple lockout/tagout situations. Where a job/task is under the control of one person, the individual employee control procedure can be used in lieu of a lockout/tagout procedure. The sample procedure can also be used as part of a complex lockout/tagout, but a more comprehensive plan should be developed, documented, and utilized.

Employees must understand the content of the company's lockout/tagout procedure before it can be effective. They must be trained to understand both the requirements and the reason for each requirement before the procedure is implemented. The following sample procedure will assist employers in developing written procedures for controlling employees' exposure to electrical energy hazards.

Lockout Procedure for ____
(Company)

    or

Tagout Procedure for ____
(Company)

■ 1.0 PURPOSE

This procedure establishes the minimum requirements for lockout (tagout) of electrical energy sources. It is to be used to ensure that conductors and circuit parts are disconnected from sources of electrical energy, locked (tagged), and tested before work begins where employees could be exposed to dangerous conditions. Sources of stored energy, such as capacitors or springs, shall be relieved of their energy. A mechanism shall be engaged to prevent re-accumulation of energy.

■ 2.0 RESPONSIBILITY

All employees shall be instructed in the safety significance of the lockout (tagout) procedure. All new or transferred employees and all other persons whose work operations are or might be in the area shall be instructed in the purpose and use of this procedure _______________________________________ [include the name(s) of person(s) or job title(s) of employees with responsibility] shall ensure that appropriate personnel receive instructions on their roles and responsibilities. All persons installing a lockout (tagout) device shall sign their names and the date on the tag.

■ 3.0 PREPARATION FOR LOCKOUT (TAGOUT)

3.1 Review drawings

Review current diagrammatic drawings (or other equally effective means), tags, labels, and signs to identify and locate all disconnecting means to determine that power is interrupted by a physical break and not a circuit interlock. Make a list of disconnecting means to be locked (tagged).

3.2 Review disconnecting means

Review disconnecting means to determine adequacy of their interrupting ability. Determine if it will be possible to verify a visible open point, or if other precautions will be necessary.

3.3 Review other work activity

Review other work activity to identify where and how other personnel might be exposed to sources of electrical energy hazards. Review other energy sources in the physical area to determine employee exposure to sources of other types of energy. Establish energy control methods for control of other hazardous energy sources in the area.

3.4 Provide voltage detector

Provide an adequately rated voltage detector to test each phase conductor or circuit part to verify that they are deenergized. (*See 12.3.*) Provide a method to determine that the voltage detector is operating satisfactorily.

3.5 Ground phase conductors or circuit parts

Where the possibility of induced voltages or stored electrical energy exists, call for grounding the phase conductors or circuit parts before touching them. Where it could be reasonably anticipated that contact with other exposed energized conductors or circuit parts is possible, call for applying ground-connecting devices.

4.0 INDIVIDUAL EMPLOYEE CONTROL PROCEDURE

The individual employee control procedure can be used when equipment with exposed conductors and circuit parts is deenergized for minor maintenance, servicing, adjusting, cleaning, inspection operating corrections, and the like. Under this procedure, the work shall be permitted to be performed without the placement of lockout/tagout devices on the disconnecting means, provided the disconnecting means is adjacent to the conductor, circuit parts, and equipment on which the work is performed. In addition, the disconnecting means must be clearly visible to all employees involved in the work, and the work must not extend beyond the work shift.

5.0 SIMPLE LOCKOUT/TAGOUT

The simple lockout/tagout procedure includes all requirements under sections 6.0 and 7.0.

6.0 SEQUENCE OF LOCKOUT (TAGOUT) SYSTEM PROCEDURES

6.1 Notify employees

Notify the employees that a lockout (tagout) system is going to be implemented and the reason therefore. The qualified employee implementing the lockout (tagout) shall know the disconnecting means location for all sources of electrical energy and the location of all sources of stored energy. The qualified person shall be knowledgeable of hazards associated with electrical energy.

6.2 Deenergize and disconnect

If the electrical supply is energized, the qualified person shall deenergize and disconnect the electric supply and relieve all stored energy.

6.3 Use lockout devices

Lock out all disconnecting means with lockout devices.

Note: *For tagout, one additional safety measure must be employed, such as opening, blocking, or removing an additional circuit element.*

6.4 Operate the disconnecting means

Attempt to operate the disconnecting means to determine that operation is prohibited.

6.5 Use a voltage-detecting instrument

Use a voltage-detecting instrument. (*See 12.3.*) Inspect the instrument for visible damage. Do not proceed if there is an indication of damage to the instrument until an undamaged device is available.

6.6 Verify before testing

Verify proper instrument operation and then test for absence of voltage.

6.7 Verify after testing

Verify proper instrument operation after testing for absence of voltage.

6.8 Install grounding equipment

Where required, install grounding equipment/conductor device on the phase conductors or circuit parts, to eliminate induced voltage or stored energy, before touching them. Where it has been determined that contact with other exposed energized conductors or circuit parts is possible, apply ground-connecting devices rated for the available fault duty.

The equipment and/or electrical source is now locked out (tagged out).

7.0 RESTORING THE EQUIPMENT AND/OR ELECTRICAL SUPPLY TO NORMAL CONDITION

7.1 Verify

After the job/task is complete, visually verify that the job/task is complete.

7.2 Remove tools

Remove all tools, equipment, and unused materials and perform appropriate housekeeping.

7.3 Remove grounding equipment

Remove all grounding equipment/conductor/devices.

7.4 Notify personnel

Notify all personnel involved with the job/task that the lockout (tagout) is complete, that the electrical supply is being restored, and to remain clear of the equipment and electrical supply.

7.5 Perform quality control

Perform any quality control tests/checks on the repaired/replaced equipment and/or electrical supply.

7.6 Remove lockout devices

Lockout devices shall be removed by the person who installed them.

7.7 Notify owner

Notify the equipment and/or electrical supply owner that the equipment and/or electrical supply is ready to be returned to normal operation.

7.8 Return to normal

Return the disconnecting means to their normal condition.

8.0 PROCEDURE INVOLVING MORE THAN ONE PERSON

8.1 Personal lockout

For a simple lockout/tagout and where more than one person is involved in the job/task, each person shall install his/her own personal lockout (tagout) device.

■ 9.0 PROCEDURE INVOLVING MORE THAN ONE SHIFT

9.1 Successive shifts

When the lockout (tagout) extends for more than one day, the lockout (tagout) shall be verified to be still in place at the beginning of the next day. Where the lockout (tagout) is continued on successive shifts, the lockout (tagout) is considered to be a complex lockout (tagout).

9.2 Communication

For complex lockout (tagout), the person-in-charge shall identify the method for transfer of the lockout (tagout) and of communication with all employees.

■ 10.0 COMPLEX LOCKOUT

10.1 Require a complex plan

A complex lockout/tagout plan is required where one or more of the following exist:

- Multiple energy sources (more than one)
- Multiple crews
- Multiple crafts
- Multiple locations
- Multiple employers
- Unique disconnecting means
- Complex or particular switching sequences
- The task continues for more than one shift and involves new workers

10.2 Prepare written plan

All complex lockout/tagout procedures shall require a written plan of execution. The plan must include the requirements in sections 6.0, 7.0, and 10.3 through 10.9.

10.3 Name person-in-charge

A person-in-charge shall be named to oversee a complex lockout/tagout procedure. At this location, ____ shall be the person-in-charge.

10.4 Assign duties

The person-in-charge shall develop a written plan of execution and communicate that plan to all persons engaged in the job/task. The person-in-charge shall be held accountable for safe execution of the complex lockout/tagout plan. The complex lockout/tagout plan must address all the concerns of employees who might be exposed, and those employees must understand how electrical energy is controlled. The person-in-charge shall ensure that each person understands the hazards to which they are exposed and the safety-related work practices they are to use.

10.5 Track personnel

All complex lockout/tagout plans identify the method to account for all persons who might be exposed to electrical hazards in the course of the lockout/tagout. Select which of the following methods is to be used:

- Each individual will install his or her own personal lockout or tagout device.
- The person-in-charge shall lock his/her key in a "lock box."

- The person-in-charge shall maintain a sign in/out log for all personnel entering the area.
- Another equally effective methodology.

10.6 Assign installation

The person-in-charge can install locks/tags or direct their installation on behalf of other employees.

10.7 Assign removal

The person-in-charge can remove locks/tags or direct their removal on behalf of other employees only after all personnel are accounted for and ensured to be clear of potential electrical hazards.

10.8 Identify transfer and communication methods (complex)

Where the complex lockout (tagout) is continued on successive shifts, the person-in-charge shall identify the method for transfer of the lockout and of communication with all employees.

10.9 Identify transfer and communication methods (simple)

For a simple lockout/tagout and where more than one person is involved in the job/task, each person shall install his/her own personal lockout (tagout) device.

11.0 DISCIPLINE

11.1 Define action for procedure violation

Knowingly violating this procedure will result in ______________________________

____

(State disciplinary actions that will be taken.)

11.2 Define action for disconnecting means violation

Knowingly operating a disconnecting means with an installed lockout device (tagout device) will result in ______________________________

____

(State disciplinary actions that will be taken.)

12.0 EQUIPMENT

12.1 Identify types of locks

Locks shall be ____

(State type and model of selected locks.)

12.2 Identify types of tags

Tags shall be ____

(State type and model to be used.)

12.3 Identify voltage-detecting device(s)

Voltage-detecting device(s) to be used shall be ______________________________

(State type and model to be used.)

■ 13.0 REVIEW

This procedure was last reviewed on ________________________________

(Provide date.)

and is scheduled to be reviewed again on ________________________________

(Not more than one year from last review.)

■ 14.0 LOCKOUT/TAGOUT TRAINING

Recommended training may include, but is not limited to, the following:

- Recognizing lockout/tagout devices
- Installing lockout/tagout devices
- Duty of employer in writing procedures
- Duty of employee in executing procedures
- Duty of person-in-charge
- Authorized and unauthorized removal of locks/tags
- Enforcing execution of lockout/tagout procedures
- Individual employee control of energy
- Simple lockout/tagout
- Complex lockout/tagout
- Using single-line and diagrammatic drawings to identify sources of energy
- Use of tags and warning signs
- Release of stored energy
- Personnel accounting methods
- Grounding needs/requirements
- Safe use of voltage-detecting instruments

Appendix **G**

Hazard/Risk Category Selections

This appendix provides information to assist in determining hazard/risk categories in conjunction with the company's electrical safety program. It contains Table 3-3.9.1 from NFPA 70E, *Standard for Electrical Safety Requirements for Employee Workplaces,* 2000 edition, that defines a hazard/risk category for each listed work task.

The hazard/risk category, in turn, defines the arc-flash protective equipment necessary for the task, as discussed in Chapters 6 and 8. The table also establishes required shock protection clothing and equipment, as discussed in Chapters 6 and 8.

The information in this table can also be used as a method of hazard analysis for common job tasks.

HAZARD/RISK CATEGORY SELECTIONS

Task (assumes equipment is energized and work is done within the flash protection boundary)	Hazard/Risk Category	Use Voltage-Rated Gloves?	Use Voltage-Rated Tools?
Panelboards Rated 240 V and Below—Notes 1 and 3			
Operate circuit breaker (CB) or fused switch with covers on	0	N	N
Operate CB or fused switch with covers off	0	N	N
Work on energized parts, including voltage testing	1	Y	Y
Remove/install CBs or fused switches	1	Y	Y
Remove bolted covers (to expose live parts)	1	N	N
Open hinged covers (to expose live parts)	0	N	N
Panelboards or Switchboards Rated 240 V to 600 V (with molded case or insulated case CBs or fused switches)—Notes 1 and 3			
Operate CB or fused switch with covers on	0	N	N
Operate CB or fused switch with covers off	1	N	N
Work on energized parts, including voltage testing	2*	Y	Y
600 V-Class Motor Control Centers (MCCs) and Busways—Note 2, except as indicated with *, and Note 3			
Operate CB or fused switch with enclosure doors closed	0	N	N
Read a panel meter while operating a meter switch	0	N	N
Operate CB or fused switch with enclosure doors open	1	N	N
Work on energized parts, including voltage testing	2*	Y	Y
Work on control circuits while working near exposed live parts 120 V or below	0	Y*	Y
Work on control circuits with exposed live parts above 120 V	2*	Y	Y
Insert or remove individual starter buckets from MCC—Note 4	3	Y	N
Apply safety grounds, after voltage test	2*	Y	N
Remove bolted covers to expose live parts	2*	N	N
Open hinged covers to expose live parts	1	N	N
600 V-Class Switchgear (with power circuit breakers or fused switches)—Notes 5 and 6			
Operate CB or fused switch with enclosure doors closed	0	N	N
Read a panel meter while operating a meter switch	0	N	N
Operate CB or fused switch with enclosure doors open	1	N	N
Work on energized parts, including voltage testing	2*	Y	Y
Work on control circuits while working near exposed live parts 120 V or below	0	Y	Y*
Work on control circuits while working near exposed live parts more than 120 V—Note 5	2*	Y	Y
Insert or remove (rack) CBs from cubicles, doors open	3	N	N
Insert or remove (rack) CBs from cubicles, doors closed	2	N	N
Apply safety grounds, after voltage test	2*	Y	N

HAZARD/RISK CATEGORY SELECTIONS

Task (assumes equipment is energized and work is done within the flash protection boundary)	Hazard/Risk Category	Use Voltage-Rated Gloves?	Use Voltage-Rated Tools?
600 V-Class Switchgear (with power circuit breakers or fused switches)—Notes 5 and 6 (continued)			
Remove bolted covers (to expose live parts)	3	N	N
Open hinged covers (to expose live parts)	2	N	N
Other Equipment, 600 V-Class (277 V through 600 V, nominal)—Note 3			
Lighting or small power transformers (600 V, maximum)			
Remove bolted covers to expose live parts	2*	N	N
Open hinged covers to expose live parts	1	N	N
Insertion or removal of revenue meter	2*	Y	N
Work on energized parts, including voltage testing—Note 5	2*	Y	Y
Apply safety grounds, after voltage test	2*	Y	N
NEMA E2 (fused contactor) motor starters, 2.3 kV through 7.2 kV			
Operate contactor with enclosure doors closed	0	N	N
Read a panel meter while operating a meter switch	0	N	N
Operate contactor with enclosure doors open	2*	N	N
Work on energized parts, including voltage testing	3	Y	Y
Work on control circuits while working near exposed live parts 120 V or below	0	Y	Y
Work on control circuits while working near live parts above 120 V	3	Y	Y
Insert or remove (rack) starters from cubicles with doors open	3	N	N
Insert or remove (rack) starters from cubicles with doors closed	2	N	N
Apply safety grounds, after voltage test	3	Y	N
Remove bolted covers (to expose live parts)	4	N	N
Open hinged covers (to expose live parts)	3	N	N
Metal-Clad Switchgear, 1 kV to 38 kV			
Operate a CB or fused switch with enclosure doors closed	2	N	N
Read a panel meter while operating a meter switch	0	N	N
Operate a CB or fused switch with enclosure doors open	4	N	N
Work on energized parts, including voltage testing	4	Y	Y
Work on control circuits while working near live parts 120V or below	2	N	N
Work on control circuits while working near live parts above 120V	4	Y	Y
Insert or remove (rack) CBs from cubicles with doors open	4	N	N
Insert or remove (rack) CBs from cubicles with doors closed	2	N	N
Apply safety grounds, after voltage test	4	Y	N
Remove bolted covers (to expose live parts)	4	N	N
Open hinged covers (to expose live parts)	3	N	N
Open voltage transformer or control power transformer compartments	4	N	N

                    (p. 2 of 3)

HAZARD/RISK CATEGORY SELECTIONS

Task (assumes equipment is energized and work is done within the flash protection boundary)	Hazard/Risk Category	Use Voltage-Rated Gloves?	Use Voltage-Rated Tools?
Other Equipment, 1 kV to 38 kV			
Operate switch, doors closed	2	N	N
Work on energized parts, including voltage testing	4	Y	Y
Remove bolted covers—up to and including 15 kV-class equipment (to expose energized parts)	4	N	N
Open hinged covers (to expose bare, energized parts)	3	N	N
Operate outdoor disconnect switch (hookstick operated)	3	Y	Y
Operate outdoor disconnect switch (gang-operated, from grade)	2	Y	Y
Examine insulated cable in manhole or other confined space	4	Y	N
Examine insulated cable in open area	2	Y	N

Legend:

Voltage-rated gloves are gloves rated and tested for the maximum line-to-line voltage upon which work will be done.

Voltage-rated tools are tools rated and tested for the maximum line-to-line voltage upon which work will be done.

2* A double-layer switching hood and hearing protection are required for this task in addition to the other hazard/risk category requirements.

Y Yes (required)

N No (not required)

Notes:

1. 25 kA short-circuit current available, 0.03 second (2-cycle) fault-clearing time.
2. 65 kA short-circuit current available, 0.03 second (2-cycle) fault-clearing time.
3. For less than 10 kA short-circuit current available, the hazard/risk category required may be reduced by one number.
4. 65 kA short-circuit current available, 0.33 second (20-cycle) fault-clearing time.
5. 65 kA short-circuit current available, up to 1.0 second (60-cycle) fault-clearing time.
6. For less than 25 kA short-circuit current available, the hazard/risk category required may be reduced by one number.

Appendix **H**

Personal Protective Equipment Matrix

This appendix contains information on personal protective equipment for use in the company's electrical safety program. It contains material from NFPA 70E, *Standard for Electrical Safety Requirements for Employee Workplaces,* 2000 edition, in the form of a matrix that can help workers select arc-flash personal protective equipment. The information in this table can be used to select PPE based on the hazard/risk categories identified in Appendix G and discussed in Chapter 6, Hazard Boundaries.

PPE MATRIX

Protective Clothing and Equipment	Protective Systems for Hazard/Risk Category				
Hazard/Risk Category Number	**0**	**1**	**2**	**3**	**4**
Non-Melting (according to ASTM F 1506-00) or Untreated Natural Fiber					
T-shirt (short-sleeved)			X	X	X
Shirt (long-sleeved)	X				
Pants (long)		X (Note 1)	X (Note 2)	X	X
FR Clothing (Note 3)					
Long-sleeved shirt		X	X	X (Note 4)	X
Pants		X (Note 1)	X (Note 2)	X (Note 4)	X
Coverall		— (Note 5)	— (Note 6)	X (Note 4)	— (Note 5)
Jacket, parka, or rainwear		AN	AN	AN	AN
FR-Protective Equipment					
Flash-suit jacket (multi-layer)					X
Flash-suit pants (multi-layer)					X
Head protection					
• Hard hat		X	X	X	X
• FR hard-hat liner				AR	AR
Eye protection					
• Safety glasses	X	X	AL	AL	AL
• Safety goggles			AL	AL	AL
Face and head area protection					
• Arc-rated face shield or flash-suit hood			X (Note 7)		X (Note 7)
• Flash-suit hood				X (Note 7)	
• Hearing protection (ear canal inserts)			AR (Note 7)	X	X
Hand protection (leather gloves—Note 8)		AN	X	X	X
Foot protection (leather work shoes)		AN	X	X	X

X = Minimum required
AN = As needed
AL = Select one in group
AR = As required

Notes:

1. Regular weight (minimum 12 oz/yd² fabric weight), untreated, denim cotton blue jeans are acceptable in lieu of FR pants. The FR pants used for Hazard/Risk Category 1 shall have a minimum arc rating of 4.

2. If the FR pants have a minimum arc-rating of 8, long pants of non-melting or untreated natural fiber are not required beneath the FR pants.

3. See NFPA 70E, Table 3-3.9.3. Arc rating for a garment is expressed in cal/cm²

4. Alternatively, two sets of FR coveralls (the inner with a minimum arc-rating of 4 and outer coverall with a minimum arc-rating of 5) over non-melting or untreated natural fiber clothing, might be used instead of FR coveralls over FR shirt and FR pants over non-melting or untreated natural-fiber clothing.

5. Alternatively, FR coveralls (minimum arc rating of 4) might be used instead of FR shirt and FR pants.

6. Alternatively, FR coveralls (minimum arc rating of 4) might be used over non-melting or untreated natural fiber pants and t-shirt.

7. Hearing protection and a face shield are required. The face shield must have a minimum arc rating of 8, with wrap-around guarding that protects the face and also the forehead, ears, and neck. (Alternatively, a flash suit hood might be used.)

8. If voltage-rated gloves are required, the leather protectors worn externally to the rubber gloves satisfy this requirement.

Appendix I

Protective Clothing Characteristics

This appendix provides information on protective clothing required in conjunction with the company's electrical safety program. Taken from NFPA 70E, *Standard for Electrical Safety Requirements for Employee Workplaces,* 2000 edition, it describes typical characteristics of protective clothing. It can be used in conjunction with Chapter 6, Hazard Boundaries, and Chapter 11, Budgeting.

The purpose of the table in this appendix is to illustrate the protective values of clothing layers. The user locates a hazard risk category in the first column and then, reading horizontally, is able to see what kind of protective clothing is typical, the weight of the material, and the protective value of the clothing. The description in the second column provides one example of clothing that could meet the requirement for the hazard/risk category. The fabric weight in the third column illustrates the protective clothing but is not a requirement. The required minimum arc rating in the fourth column provides a correlation between the hazard/risk category and the required clothing.

The hazard/risk category is determined by the hazard/risk analysis—when the flash-hazard analysis is performed or when determined from Appendix G. Note that Appendix G provides default hazard/risk category information.

PROTECTIVE CLOTHING CHARACTERISTICS

Typical Protective Clothing Systems

Hazard Risk Category	Clothing Description (Typical number of clothing layers is given in parentheses.)	Total Weight oz/yd²	Required Minimum Arc Rating[a] of PPE in cal/cm²
0	Non-melting, flammable materials (i.e., untreated cotton, wool, rayon, or silk, or blends of these materials) with a fabric weight of at least 4.5 oz/yd² (1)	4.5–14	For use at up to 2 cal/cm² incident energy exposure, but materials have no arc rating
1	Flame-resistant (FR) shirt and FR pants *or* FR coverall (1)	4.5–12	4
2	Cotton underwear—conventional short sleeves and briefs/shorts, plus FR shirt and FR pants (2)	9–16	8
3	Cotton underwear plus FR shirt and FR pants plus FR coverall, *or* cotton underwear plus two FR coveralls (3)	16–20	25
4	Cotton underwear plus FR shirt and FR pants, plus multi-layer flash suit (3 or more)	24–30	40[b]

[a]Arc rating is defined in Part I, Section I-2, and can be either ATPV or E_{BT}. ATPV is defined in ASTM F 1959-99 as the incident energy on a fabric or material that results in sufficient heat transfer through the fabric or materials to cause the onset of a second-degree burn based on the Stoll curve. E_{BT} is defined in ASTM F 1959-99 as the average of the five highest incident energy exposure values below the Stoll curve where the specimens do not exhibit break-open. E_{BT} is reported when ATPV cannot be measured, due to FR fabric break-open.

[b]If the calories are more than 40, then the work must be done de-energized.

Appendix J

Simplified Protective Clothing Table

This appendix provides a simplified system for selecting protective clothing needed in the company's electrical safety program that can be used in conjunction with Chapter 6, Hazard Boundaries, Chapter 8, Personal Protective Equipment, and Chapter 12, Program Information Administration. It contains Table F.1 from NFPA 70E, *Standard for Electrical Safety Requirements for Employee Workplaces,* 2000 edition, which describes a two-stage protective clothing selection system.

The table illustrates that an employer can select a protective clothing scheme that is relatively simple to administer. This scheme requires workers to use Category 2 protective clothing for all hazard/risk Category 1 and 2 work tasks. The table also requires Category 4 protective equipment for all Category 3 and 4 work tasks. The advantage of the simplified protective clothing table is that administration of the protective clothing procedure is relatively easy. The obvious disadvantage is that the workers are expected to wear more protective clothing than might be necessary.

SIMPLIFIED PROTECTIVE CLOTHING TABLE

Clothing^(See Note)	Applicable Tasks
Everyday Work Clothing	All Hazard/Risk Category 1 and 2 tasks listed in NFPA 70E Table 3-3.9.1 [Appendix G of this book].
Fire resistant (FR) long-sleeve shirt [minimum arc-thermal protective value (ATPV) rating of 4] worn over an untreated cotton t-shirt with FR pants (minimum ATPV rating of 8) *Or* FR coveralls (minimum ATPV rating of 4) worn over an untreated cotton t-shirt (or an untreated natural fiber long-sleeved shirt) with untreated natural fiber pants	On systems operating at less than 1000 volts, these tasks include work on all equipment *except* on the following: • Insertion or removal of low-voltage motor starter "buckets" • Insertion or removal of power circuit breakers from switchgear cubicles • Removal of bolted covers from switchgear On systems operating at 1000 volts or greater, tasks also include the operation of switching devices with equipment enclosure doors closed.
Electrical "Switching" Clothing	All hazard/risk category 3 and 4 tasks listed in NFPA 70E Table 3-3.9.1 [Appendix G of this book].
Multi-layer FR flash jacket and FR bib overalls worn over either FR coveralls (minimum ATPV rating of 4) or FR long-sleeve shirt and FR pants (minimum ATPV rating of 4), worn over untreated natural fiber long-sleeve shirt and pants, worn over an untreated cotton t-shirt Insulated FR coveralls (with a minimum ATPV rating of 25, independent of other layers) worn over untreated natural fiber long-sleeved shirt with untreated denim cotton blue jeans ("regular weight," minimum 12-oz/yd^2 fabric weight), worn over an untreated cotton t-shirt.	On systems operating at 1000 volts or greater, these tasks include work on exposed live parts of all equipment. On systems of less than 1000 volts, tasks include insertion or removal of low-voltage motor starter MCC "buckets," insertion or removal of plug-in devices into or from busway, insertion or removal of plug-in devices into or from busway, insertion or removal of power circuit breakers, and removal of bolter covers from switchgear.

Note   *Other PPE is required for the specific tasks listed in NFPA 70E Tables 3-3.9.1 and 3-3.9.2, which include arc-rated face shields or flash suit hoods, FR hard hat liners, safety glasses or safety goggles, hard hat, hearing protection, leather gloves, voltage-rated gloves, and voltage-rated tools.*

Appendix K
Budget Samples and Checklist

This appendix contains two sample budgets for operating an electrical safety program, as discussed in Chapter 3, Setting Up the Electrical Safety Program, and Chapter 11, Budgeting. The budgets define the costs associated with setting up and administering the program, such as building space, furniture, computers and software, copies and files, and so forth. These sample budgets, however, do not include costs associated with generating and maintaining procedures that are, or may be, necessary.

The first example is based on a hypothetical manufacturing facility with an established electrical safety program. The example provides a general idea of the resources that might be required for a company of the size indicated at the beginning of the budget to fund an established electrical safety program within its overall safety program. The second example is based on a facility the same size as the first facility. However, in this example, the company does not have an existing electrical safety program. In the startup budget, the company must completely fund resources for each required line item in the budget.

■ SAMPLE BUDGET OUTLINE FOR AN ESTABLISHED PROGRAM

Company A is the manufacturer of computerized equipment that employs 1,200 workers. The company has three buildings, comprising 750,000 square feet. The buildings are located in a campus-style configuration. The company has an established Electrical Safety Program that is in compliance with OSHA regulations and incorporates best work practices. The company has assigned the facilities department the responsibility of incorporating the electrical safety program into their overall safety program. The facilities manager has assigned this task to the maintenance manager, who is also in charge of the overall safety program. This task is in addition to all of his maintenance responsibilities.

The maintenance manager's professional background includes trade school and work for an HVAC installation as a service contractor, HVAC maintenance mechanic, and maintenance supervisor. He has limited electrical experience, other than that required for HVAC service work. He has basic computer skills. He is proficient in formulating budgets, but he has limited experience developing presentations using computer programs such as Microsoft® PowerPoint®. He has provided most of the on-site training to the maintenance staff. He uses outside contractors to complement and support the maintenance staff.

The facility has a staff of maintenance personnel, including electricians, HVAC mechanics, carpenters, laborers, and custodians. The group performs the majority of the equipment and system maintenance. Outside contractors provide the resource for new installations and major remodeling in the facility. The contractors also work per diem for the maintenance department to supplement the group.

Table K-1 illustrates the firm's budget for funding its established electrical safety program within its overall safety program.

TABLE K-1   SAMPLE ELECTRICAL SAFETY PROGRAM BUDGET FOR AN ESTABLISHED PROGRAM

Section	Description	Capital	Expense
1	**Staff Costs**		
(1A)	Manager, full time or pro-rated[a]		7,000
(1B)	Assistant or administrative, full time or pro-rated		8,256
	Staff Subtotal		**15,256**
2	**Facilities Expense**		
(2A)	Additional space required		0
(2B)	Remodeling		0
(2C)	Furniture	500	
(2D)	Computer equipment		800
(2E)	Telephone(s)		0
(2F)	Internet service		600
(2G)	Reference materials		200
	Facilities Expense Subtotal		**2,100**
3	**Program Development Costs**		
(3A)	Update reference drawings		2,000
(3B)	Site assessment		1,000
(3C)	Hazard/risk analysis		1,000
(3D)	Task assessment		400
(3E)	Safety program book		0
(3F)	Employees and contractors book		0
	Program Developmental Subtotal		**4,400**
4	**Training**		
(4A)	Manager training		2,000
(4B)	Staff training, on site		1,000
(4C)	Staff training, off site		4,000
(4D)	Employee training		500
(4E)	Contractor training		500
	Training Subtotal		**8,000**
5	**Equipment**		
(5A)	Personal protective equipment		1,000
(5B)	Lockout/tagout equipment		200
(5C)	Other equipment		0
	Equipment Subtotal		**1,200**
6	**Administration**		
(6A)	Auditing costs		0
(6B)	Recordkeeping costs		0
(6C)	Update program book		500

(6D)	Training program		3,000
(6E)	Employee awareness program		2,000
	Administration Subtotal		**5,500**
	Capital and Expense Totals	**500**	**36,456**
	Total Budget		**36,956**

[a]Base salary times 10 percent

■ SAMPLE BUDGET OUTLINE FOR A START-UP PROGRAM

Company B is the manufacturer of computerized equipment that employs 1,200 workers. The company has three buildings, comprising 750,000 square feet. The buildings are located in a campus-style configuration. The company plans to develop an Electrical Safety Program that is in compliance with OSHA regulations and incorporates best work practices. The company has assigned the facilities department the responsibility of incorporating the electrical safety program into their overall safety program. The facilities manager has assigned this task to the maintenance manager, who is also in charge of the overall safety program. This task is in addition to all of his maintenance responsibilities.

The maintenance manager's professional background includes trade school and work for an HVAC installation as a service contractor, HVAC maintenance mechanic, and maintenance supervisor. He has limited electrical experience, other than that required for HVAC service work. He has basic computer skills. He is proficient in formulating budgets, but he has limited experience developing presentations using computer programs such as Microsoft® PowerPoint®. He has provided most of the on-site training to the maintenance staff. He uses outside contractors to complement and support the maintenance staff.

The facility has a staff of maintenance personnel, including electricians, HVAC mechanics, carpenters, laborers, and custodians. The group performs the majority of the equipment and system maintenance. Outside contractors provide the resource for new installations and major remodeling in the facility. The contractors also work per diem for the maintenance department to supplement the group.

Table K-2 illustrates the firm's budget for starting up an electrical safety program within its overall safety program. Once established, annual costs will be lower than those illustrated in Table K-2.

TABLE K-2   SAMPLE ELECTRICAL SAFETY PROGRAM BUDGET FOR A START-UP PROGRAM

Section	Description	Capital	Expense
1	**Staff Costs**		
(1A)	Manager, full time or pro-rated		7,000
(1B)	Assistant or administrative, full time or pro-rated		22,200
	Staff Subtotal		**29,200**
2	**Facilities Expense**		
(2A)	Additional space required		0
(2B)	Remodeling		0
(2C)	Furniture	800	

(continued)

TABLE K-2   *(Continued)*

Section	Description	Capital	Expense
2	**Facilities Expense** *(continued)*		
(2D)	Computer equipment		800
(2E)	Telephone(s)		0
(2F)	Internet service		600
(2G)	Reference materials		200
	Facilities Expense Subtotal		**2,400**
3	**Program Development Costs**		
(3A)	Update reference drawings		4,000
(3B)	Site assessment		1,000
(3C)	Hazard/risk analysis		1,000
(3D)	Task assessment		400
(3E)	Safety program book		200
(3F)	Employees and contractors book		500
	Program Developmental Subtotal		**7,100**
4	**Training**		
(4A)	Manager training		4,000
(4B)	Staff training, on site		2,000
(4C)	Staff training, off site		10,000
(4D)	Employee training		500
(4E)	Contractor training		500
	Training Subtotal		**17,000**
5	**Equipment**		
(5A)	Personal protective equipment		3,000
(5B)	Lockout/tagout equipment		1,000
(5C)	Other equipment		1,000
	Equipment Subtotal		**5,000**
6	**Administration**		
(6A)	Auditing costs		0
(6B)	Recordkeeping costs		0
(6C)	Update program book		500
(6D)	Training program		1,000
(6E)	Employee awareness program		2,000
	Administration Subtotal		**3,500**
	Capital and Expense Totals	800	**64,200**
	Total Budget		**65,000**

Glossary of Terms

Many of these terms are extracted from NFPA 70E-2004 and are intended to provide consistency and clarity for developing an electrical safety program.

Accessible (as applied to wiring methods). Capable of being removed or exposed without damaging the building structure or finish or not permanently closed in by the structure or finish of the building.

Accessible (as applied to equipment). Admitting close approach; not guarded by locked doors, elevation, or other effective means.

Accessible, Readily (Readily Accessible). Capable of being reached quickly for operation, renewal, or inspections without requiring those to whom ready access is requisite to climb over or remove obstacles or to resort to portable ladders, and so forth.

Ampacity. The current, in amperes, that a conductor can carry continuously under the conditions of use without exceeding its temperature rating.

Appliance. Utilization equipment, generally other than industrial, that is normally built in standardized sizes or types and is installed or connected as a unit to perform one or more functions such as clothes washing, air conditioning, food mixing, deep frying, and so forth.

Approved. The National Fire Protection Association does not approve, inspect, or certify any installations, procedures, equipment, or materials; nor does it approve or evaluate testing laboratories. In determining the acceptability of installations, procedures, equipment, or materials, the authority having jurisdiction may base acceptance on compliance with NFPA or other appropriate standards. In the absence of such standards, said authority may require evidence of proper installation, procedure, or use. The authority having jurisdiction may also refer to the listings or labeling practices of an organization that is concerned with product evaluations and is thus in a position to determine compliance with appropriate standards for the current production of listed items.

Arc-Over Distance. The physical dimension at which the difference of potential overcomes the ability of air to resist the flow of current from one electrode to another under the existing environmental conditions (the basic minimum air insulation distance).

Arc Rating. The maximum incident energy resistance demonstrated by a material (or a layered system of materials) prior to breakopen or at the onset of a second-degree skin burn. Arc rating is normally expressed in cal/cm^2.

Note: *"Breakopen" is a material response evidenced by the formation of one or more holes in the innermost layer of flame-resistant material that would allow flame to pass through the material.*

Armored Cable. A fabricated assembly of insulated conductors in a metallic enclosure.

Attachment Plug (Plug Cap) (Plug).   A device that, by insertion in a receptacle, establishes a connection between the conductors of the attached flexible cord and the conductors connected permanently to the receptacle.

Automatic.   Self-acting, operating by its own mechanism when actuated by some impersonal influence, as, for example, a change in current, pressure, temperature, or mechanical configuration.

Bare Hand Work.   A technique of performing work on live parts, after the employee has been raised to the potential of the live part.

Barricade.   A physical obstruction such as tapes, cones, or A-frame-type wood or metal structures intended to provide a warning about and to limit access to a hazardous area.

Barrier.   A physical obstruction that is intended to prevent contact with equipment or live parts or to prevent unauthorized access to a work area.

Bathroom.   An area including a basin with one or more of the following: a toilet, a tub, or a shower.

Bonding (Bonded).   The permanent joining of metallic parts to form an electrically conductive path that ensures electrical continuity and the capacity to conduct safely any current likely to be imposed.

Bonding Jumper.   A reliable conductor to ensure the required electrical conductivity between metal parts required to be electrically connected.

Branch Circuit.   The circuit conductors between the final overcurrent device protecting the circuit and the outlet(s).

Building.   A structure that stands alone or that is cut off from adjoining structures by fire walls with all openings therein protected by approved fire doors.

Cabinet.   An enclosure that is designed for either surface mounting or flush mounting and is provided with a frame, mat, or trim in which a swinging door or doors are or can be hung.

Cable Tray System.   A unit or assembly of units or sections and associated fittings forming a rigid structural system used to securely fasten or support cables and raceways.

Cablebus.   An assembly of insulated conductors with fittings and conductor terminations in a completely enclosed, ventilated protective metal housing. Cablebus is ordinarily assembled at the point of installation from the components furnished or specified by the manufacturer in accordance with instructions for the specific job. This assembly is designed to carry fault current and to withstand the magnetic forces of such current.

Circuit Breaker.   A device designed to open and close a circuit by nonautomatic means and to open the circuit automatically on a predetermined overcurrent without damage to itself when properly applied within its rating.

Class I locations. Class I locations are those in which flammable gases or vapors are or may be present in the air in quantities sufficient to produce explosive or ignitible mixtures. Class I locations shall include those specified in (1) or (2).

(1) **Class I, Division 1.**   A Class I, Division 1 location is a location:

(1) In which ignitible concentrations of flammable gases or vapors can exist under normal operating conditions, or

(2) In which ignitible concentrations of such gases or vapors may exist frequently because of repair or maintenance operations or because of leakage, or

(3) In which breakdown or faulty operation of equipment or processes might release ignitible concentrations of flammable gases or vapors and might also cause simultaneous failure of electrical equipment in such a way as to directly cause the electrical equipment to become a source of ignition.

Note 1: This classification usually includes the following locations:

(1) Where volatile flammable liquids or liquefied flammable gases are transferred from one container to another

(2) Interiors of spray booths and areas in the vicinity of spraying and painting operations where volatile flammable solvents are used

(3) Locations containing open tanks or vats of volatile flammable liquids

(4) Drying rooms or compartments for the evaporation of flammable solvents

(5) Locations containing fat- and oil-extraction equipment using volatile flammable solvents

(6) Portions of cleaning and dyeing plants where flammable liquids are used

(7) Gas generator rooms and other portions of gas manufacturing plants where flammable gas may escape

(8) Inadequately ventilated pump rooms for flammable gas or for volatile flammable liquids

(9) The interiors of refrigerators and freezers in which volatile flammable materials are stored in open, lightly stoppered, or easily ruptured containers

(10) All other locations where ignitible concentrations of flammable vapors or gases are likely to occur in the course of normal operations

Note 2: In some Division 1 locations, ignitible concentrations of flammable gases or vapors could be present continuously or for long periods of time. Examples include the following:

(1) The inside of inadequately vented enclosures containing instruments normally venting flammable gases or vapors to the interior of the enclosure

(2) The inside of vented tanks containing volatile flammable liquids

(3) The area between the inner and outer roof sections of a floating roof tank containing volatile flammable fluids

(4) Inadequately ventilated areas within spraying or coating operations using volatile flammable fluids

(5) The interior of an exhaust duct that is used to vent ignitible concentrations of gases or vapors

Experience has demonstrated the prudence of avoiding the installation of instrumentation or other electric equipment in these particular areas altogether or where it cannot be avoided because it is essential to the process and other locations are not feasible using electric equipment or instrumentation approved for the specific application or consisting of intrinsically safe systems.

(2) **Class I, Division 2.**   A Class I, Division 2 location is a location

(1) In which volatile flammable liquids or flammable gases are handled, processed, or used, but in which the liquids, vapors, or gases will normally be confined

within closed containers or closed systems from which they can escape only in case of accidental rupture or breakdown of such containers or systems or in case of abnormal operation of equipment, or

(2) In which ignitible concentrations of gases or vapors are normally prevented by positive mechanical ventilation, and which might become hazardous through failure or abnormal operation of the ventilating equipment, or

(3) That is adjacent to a Class I, Division 1 location, and to which ignitable concentrations of gases or vapors might occasionally be communicated unless such communication is prevented by adequate positive-pressure ventilation from a source of clean air and effective safeguards against ventilation failure are provided.

Note 1: This classification usually includes locations where volatile flammable liquids or flammable gases or vapors are used but that, in the judgment of the authority having jurisdiction, would become hazardous only in case of an accident or of some unusual operating condition. The quantity of flammable material that might escape in case of accident, the adequacy of ventilating equipment, the total are involved, and the record of the industry or business with respect to explosions or fires are all factors that merit consideration in determining the classification and extent of each location.

Note 2: Piping without valves, checks, meters, and similar devices would not ordinarily introduce a hazardous condition even though used for flammable liquids or gases. Depending on factors such as the quantity and size of the containers and ventilation, locations used for the storage of flammable liquids or liquefied or compressed gases in sealed containers may be considered either hazardous (classified) or unclassified locations. See NFPA 30-2000, Flammable and Combustible Liquids Code, *and NFPA 58-2001,* Liquefied Petroleum Gas Code.

Class I, Zone 0, 1, and 2 locations.   Class I, Zone 0, 1, and 2 locations are those in which flammable gases or vapors are or may be present in the air in quantities sufficient to produce explosive or ignitible mixtures. Class I, Zone 0, 1, and 2 location shall include those specified in (1), (2), and (3).

(1) **Class I, Zone 0.** A Class I, Zone 0 location is a location in which

(1) Ignitible concentrations of flammable gases or vapors are present continuously, or

(2) Ignitible concentrations of flammable gases or vapors are present for long periods of time.

Note 1: As a guide in determining when flammable gases or vapors are present continuously or for long periods of time, refer to ANSI/API RP 505-1997, Recommended Practice for Classification of Locations for Electrical Installations of Petroleum Facilities Classified as Class I, Zone 0, Zone 1, or Zone 2; *ISA 12.24.01-1998,* Recommended Practice for Classification of Locations for Electrical Installations Classified as Class I, Zone 0, Zone 1, or Zone 2; *IEC 60079-10-1995,* Electrical Apparatus for Explosive Gas Atmospheres, Classifications of Hazardous Areas; *and* Area Classification Code for Petroleum Installations, Model Code, *Part 15, Institute of Petroleum.*

Note 2: This classification includes locations inside vented tanks or vessels that contain volatile flammable liquids; inside inadequately vented spraying or coating enclosures, where volatile flammable solvents are used; between the inner and outer roof sections of a floating roof tank containing volatile flammable liquids; inside open vessels, tanks and pits containing volatile flammable liquids; the interior of an exhaust duct that is used to vent ignitible concentrations of gases or vapors; and inside inadequately ventilated enclosures that contain normally venting instruments utilizing or analyzing flammable fluids and venting to the inside of the enclosures.

Note 3: It is not good practice to install electrical equipment in Zone 0 locations except when the equipment is essential to the process or when other locations are not feasible. [See Note 2] If it is necessary to install electrical systems in a Zone 0 location, it is good practice to install intrinsically safe systems.

(2) **Class I, Zone 1.** A Class I, Zone 1 location is a location

(1) In which ignitible concentrations of flammable gases or vapors are likely to exist under normal operating conditions; or

(2) In which ignitible concentrations of flammable gases or vapors may exist frequently because of repair or maintenance operations or because of leakage; or

(3) In which equipment is operated or processes are carried on, of such a nature that equipment breakdown or faulty operations could result in the release of ignitible concentrations of flammable gases or vapors and also cause simultaneous failure of electrical equipment in a mode to cause the electrical equipment to become a source of ignition; or

(4) That is adjacent to a Class I, Zone 0 location from which ignitible concentrations of vapors could be communicated, unless communication is prevented by adequate positive pressure ventilation from a source of clean air and effective safeguards against ventilation failure are provided.

Note 1: Normal operation is considered the situation when plant equipment is operating within its design parameters. Minor releases of flammable material could be part of normal operations. Minor releases include the releases from mechanical packings on pumps. Failures that involve repair or shutdown (such as the breakdown of pump seals and flange gaskets, and spillage caused by accidents) are not considered normal operation.

Note 2: This classification usually includes locations where volatile flammable liquids or liquefied flammable gases are transferred from one container to another. In areas in the vicinity of spraying and painting operations where flammable solvents are used; adequately ventilated drying rooms or compartments for evaporation of flammable solvents; adequately ventilated locations containing fat and oil extraction equipment using volatile flammable solvents; portions of cleaning and dyeing plants where volatile flammable liquids are used; adequately ventilated gas generator rooms and other portions of gas manufacturing plants where flammable gas could escape; inadequately ventilated pump rooms for flammable gas or for volatile flammable liquids; the interiors of refrigerators or freezers in which volatile flammable materials are stored in the open, lightly stoppered, or in easily ruptured containers; and other locations where ignitible concentrations of flammable vapors or gases are likely to occur in the course of normal operation but not classified Zone 0.

(3) **Class I, Zone 2.** A Class I, Zone 2 location is a location

(1) In which ignitible concentrations of flammable gases or vapors are not likely to occur in normal operation and, if they do occur, will exist only for a short period; or

(2) In which volatile flammable liquids, flammable gases, or flammable vapors are handled, processed, or used but in which the liquids, gases, or vapors normally are confined within closed containers of closed containers of closed systems from which they can escape, only as a result of accidental rupture or breakdown of the containers or system, or as a result of the abnormal operation of the equipment with which the liquids or gases are handled, processed, or used; or

(3) In which ignitible concentrations of flammable gases or vapors normally are prevented by positive mechanical ventilation but which may become hazardous as a result of failure or abnormal operation of the ventilating equipment; or

(4) That is adjacent to a Class I, Zone 1 location, from which ignitible concentrations of flammable gases or vapors could be communicated, unless such communication is prevented by adequate positive-pressure ventilation from a source of clean air and effective safeguards against ventilation failure are provided.

Note: *The Zone 2 classification usually includes locations where volatile flammable liquids or flammable gases or vapors are used but which would become hazardous only in case of an accident or of some unusual operating condition.*

Class II Locations. Class II locations are those that are hazardous because of the presence of combustible dust. Class II locations shall include those in (1) and (2).

(1) **Class II, Division 1.** A Class II, Division 1 location is a location

(1) In which combustible dust is in the air under normal operating conditions in quantities sufficient to produce explosive or ignitible mixtures, or

(2) Where mechanical failure or abnormal operation of machinery or equipment might cause such explosive or ignitible mixtures to be produced, and might also provide a source of ignition through simultaneous failure of electric equipment, through operation of protection devices, or from other causes, or

(3) In which combustible dusts of an electrically conductive nature may be present in hazardous quantities.

Note: *Combustible dusts that are electrically nonconductive include dusts produced in the handling and processing of grain and grain products, pulverized sugar and cocoa, dried egg and milk powders, pulverized spices, starch and pasts, potato and wood-flour, oil meal from beans and seed, dried hay, and other organic materials that could produce combustible dusts when processed or handled. Only Group E dusts are considered to be electrically conductive for classification purposes. Dusts containing magnesium or aluminum are particularly hazardous, and the use of extreme precaution is necessary to avoid ignition and explosion.*

(2) **Class II, Division 2.**   A Class II, Division 2 location is a location

(1) Where combustible dust is not normally in the air in quantities sufficient to produce explosive or ignitible mixtures, and dust accumulations are normally insufficient to interfere with the normal operation of electrical equipment or other apparatus, but combustible dust may be in suspension in the air as a result of infrequent malfunctioning of handling or processing equipment and

(2) Where combustible dust accumulations on, in, or in the vicinity of the electrical equipment may be sufficient to interfere with the safe dissipation of heat from electrical equipment or may be ignitible by abnormal operation or failure of electrical equipment.

Note 1:   The quantity of combustible dust that may be present and the adequacy of dust removal systems are factors that merit consideration in determining the classification and may result in an unclassified area.

Note 2:   Where products such as seed are handled in manner that produces low quantities of dust, the amount of dust deposited could not warrant classification.

Class III Locations.   Class III locations are those that are hazardous because of the presence of easily ignitible fibers or flyings, but in which such fibers or flyings are not likely to be in suspension in the air in quantities sufficient to produce ignitible mixtures. Class III locations shall include those specified in (1) and (2).

(1)   **Class III, Division 1.**   A Class III, Division 1 location is a location in which easily ignitible fibers or materials producing combustible flyings are handled, manufactured, or used.

Note 1:   Such locations usually include some parts of rayon, cotton, and other textile mills; combustible fiber manufacturing and processing plants; cotton gins and cotton-seed mills; flax-processing plants; clothing manufacturing plants; woodworking plants; and establishments and industries involving similar hazardous processes or conditions.

Note 2:   Easily ignitible fibers or flyings include rayon, cotton (including cotton linters and cotton waste), sisal or henequen, istle, jute, hemp, tow, cocoa fiber, oakum, baled waste kapok, Spanish moss, excelsior, and other materials of similar nature.

(2)   **Class III, Division 2.**   A Class III, Division 2 location is a location in which easily ignitible fibers are stored or handled other than in the process of manufacture.

Concealed.   Rendered inaccessible by the structure or finish of the building. Wires in concealed raceways are considered concealed, even though they may become accessible by withdrawing them.

Conductive.   Suitable for carrying electric current.

Conductor, Bare.   A conductor having no covering or electrical insulation whatsoever.

Conductor, Covered.   A conductor encased within material of composition or thickness that is not recognized by this standard as electrical insulation.

Conductor, Insulated.   A conductor encased within material of composition and thickness that is recognized by this standard as electrical insulation.

Conduit Body.   A separate portion of a conduit or tubing system that provides access through a removable cover(s) to the interior of the system at a junction of two or more sections of the system or at a terminal point of the system.

Note: Boxes such as FS and FD or larger cast or sheet metal boxes are not classified as conduit bodies.

Controller.   A device or group of devices that serves to govern, in some predetermined manner, the electric power delivered to the apparatus to which it is connected.

Cooking Unit, Counter-Mounted.   A cooking appliance designed for mounting in or on a counter and consisting of one or more heating elements, internal wiring, and built-in or mountable controls.

Cutout Box.   An enclosure designed for surface mounting that has swinging doors or covers secured directly to and telescoping with the walls of the box proper.

Dead Front.   Without live parts exposed to a person on the operating side of the equipment.

Deenergized.   Free from any electrical connection to a source of potential difference and from electrical charge; not having a potential different from that of the earth.

Device.   A unit of an electrical system that is intended to carry but not utilize electric energy.

Dielectric Heating.   Heating of a nominally insulating material due to its own dielectric losses when the material is placed in a varying electric field.

Disconnecting Means.   A device, or group of devices, or other means by which the conductors of a circuit can be disconnected from their source of supply.

Effective Ground-Fault Current Path.   An intentionally constructed, permanent, low-impedance electrically conductive path designed and intended to carry current under ground-fault conditions from the point of a ground-fault on a wiring system to the electrical supply source.

Electric Sign.   A fixed, stationary, or portable self-contained, electrically illuminated utilization equipment with words or symbols designed to convey information or attract attention.

Electrical Hazard.   A dangerous condition such that contact or equipment failure can result in electric shock, arc flash burn, thermal burn, or blast.

Note: Class 2 power supplies, listed low voltage lighting systems, and similar sources are examples of circuits or systems that are not considered an electrical hazard.

Electrical Safety.   Recognizing hazards associated with the use of electrical energy and taking precautions so that hazards do not cause injury or death.

Electrical Single-Line Diagram.   A diagram that shows, by means of single lines and graphic symbols, the course of an electric circuit or system of circuits and the component devices or parts used in the circuit or system.

Electrically Safe Work Condition.   A state in which the conductor or circuit part to be worked on or near has been disconnected from energized parts, locked/tagged in accordance with established standards, tested to ensure the absence of voltage, and grounded if determined necessary.

Enclosed.   Surrounded by a case, housing, fence, or wall(*s*) that prevent*s* persons from accidentally contacting energized parts.

Enclosure.   The case or housing of apparatus, or the fence or walls surrounding an installation to prevent personnel from accidentally contacting energized parts, or to protect the equipment from physical damage.

Energized.   Electrically connected to or having a source of voltage.

Equipment.   A general term including material, fittings, devices, appliances, *luminaries (fixtures),* apparatus, and the like used as a part of, or in connection with, an electrical installation.

Explosionproof Apparatus.   Apparatus enclosed in a case that is capable of withstanding an explosion of a specified gas or vapor that may occur within it and of preventing the ignition of a specified gas or vapor surrounding the enclosure by sparks, flashes, or explosion of the gas or vapor within, and that operates at such an external temperature that a surrounding flammable atmosphere will not be ignited thereby.

Exposed. (as applied to live parts.)   Capable of being inadvertently touched or approached nearer than a safe distance by a person. It is applied to parts that are not suitably guarded, isolated, or insulated.

Exposed. (as applied to wiring methods.)   On or attached to the surface or behind panels designed to allow access.

Exposed.   A circuit is in such a position that, in case of failure of supports or insulation, contact with another circuit may result.

Externally Operable.   Capable of being operated without exposing the operator to contact with live parts.

Feeder.   All circuit conductors between the service equipment, the source of a separately derived system, or other power supply source and the final branch-circuit overcurrent device.

Fitting.   An accessory such as a locknut, bushing, or other part of a wiring system that is intended primarily to perform a mechanical rather than an electrical function.

Flame-Resistant (FR).   The property of a material whereby combustion is prevented, terminated, or inhibited following the application of a flaming or non-flaming source of ignition, with or without subsequent removal of the ignition source.

Note: Flame resistance can be an inherent property of a material, or it can be imparted by a specific treatment applied to the material.

Flash Hazard.   A dangerous condition associated with the release of energy caused by an electric arc.

Flash Hazard Analysis.   A study investigating a worker's potential exposure to arc-flash energy, conducted for the purpose of injury prevention and the determination of safe work practices and the appropriate levels of PPE.

Flash Protection Boundary.   An approach limit at a distance from exposed live parts within which a person could receive a second degree burn if an electrical arc flash were to occur.

Flash Suit.   A complete FR clothing and equipment system that covers the entire body, except for the hands and feet. This includes pants, jacket, and bee-keeper type hood fitted with a face shield.

Fuse.   An overcurrent protective device with a circuit-opening fusible part that is heated and severed by the passage of overcurrent through it.

Note: A fuse comprises all the parts that form a unit capable of performing the prescribed functions. It may or may not be the complete device necessary to connect it into an electrical circuit.

Ground.   A conducting connection, whether intentional or accidental, between an electrical circuit or equipment and the earth or to some conducting body that serves in place of the earth.

Grounded.   Connected to earth or to some conducting body that serves in place of the earth.

Grounded Conductor.   A system or circuit conductor that is intentionally grounded.

Grounded, Effectively.   Intentionally connected to earth through a ground connection or connections of sufficiently low impedance and having sufficient current-carrying capacity to prevent the buildup of voltages that may result in undue hazards to connected equipment or to persons.

Grounding Conductor.   A conductor used to connect equipment or the grounded circuit of a wiring system to a grounding electrode or electrodes.

Grounding Conductor, Equipment.   The conductor used to connect the noncurrent-carrying metal parts of equipment, raceways, and other enclosures to the system grounded conductor, the grounding electrode conductor, or both, at the service equipment or at the source of a separately derived system.

Grounding Electrode Conductor.   The conductor used to connect the grounding electrode(s) to the equipment grounding conductor, to the grounded conductor, or to both, at each service, at each building or structure where supplied from a common service, or at the source of a separately derived system.

Ground Fault.   An unintentional, electrically conducting connection between an ungrounded conductor of an electrical circuit and the normally non-current-carrying conductors, metallic enclosures, metallic raceways, metallic equipment, or earth.

Ground-Fault Circuit-Interrupter.   A device intended for the protection of personnel that functions to de-energize a circuit or portion thereof within an established period of time when a current to ground exceeds the values established for a Class A device.

Note: Class A ground-fault circuit-interrupters trip when the current to ground has a value in the range of 4 mA to 6 mA. For further information. See UL 943, Standard for Ground-Fault Circuit-Interrupters.

Ground-Fault Current Path.   An electrically conductive path from the point of a ground fault on a wiring system through normally non-current-carrying conductors, equipment, or the earth to the electrical supply source.

Note: Examples of ground-fault current paths could consist of any combination of equipment grounding conductors, metallic raceways, metallic cable sheaths, electrical equipment, and any other electrically conductive material such as metal water and gas piping, steel framing members, stucco mesh, metal ducting, reinforcing steel, shields of communications cables, and the earth itself.

Guarded.   Covered, shielded, fenced, enclosed, or otherwise protected by means of suitable covers, casings, barriers, rails, screens, mats, or platforms to remove the likelihood of approach or contact by persons or objects to a point of danger.

Health Care Facilities.   Buildings or portions of buildings in which medical, dental, psychiatric, nursing, obstetrical, or surgical care are provided. Health care facilities include, but are not limited to, hospitals, nursing homes, limited care facilities, clinics, medical and dental offices, and ambulatory care centers, whether permanent or movable.

Heating Equipment.   Any equipment used for heating purposes whose heat is generated by induction or dielectric methods.

Hoistway.   Any shaftway, hatchway, well hole, or other vertical opening or space in which an elevator or dumbwaiter is designed to operate.

Identified. (as applied to equipment.) Recognizable as suitable for the specific purpose, function, use, environment, application and so forth, where described in a particular code or standard requirement.

Note: Some examples of ways to determine suitability of equipment for a specific purpose, environment, or application include investigations by a qualified testing laboratory (listing and labeling), an inspection agency, or other organizations concerned with product evaluation.

Incident Energy. The amount of energy impressed on a surface, a certain distance from the source, generated during an electrical arc event. One of the units used to measure incident energy is calories per centimeter squared (cal/cm^2).

Induction Heating. The heating of a nominally conductive material due to its own I^2R losses when the material is placed in a varying electromagnetic field.

Insulated. Separated from other conducting surfaces by a dielectric (including air space) offering a high resistance to the passage of current.

Note: When an object is said to be insulated, it is understood to be insulated for the conditions to which it is normally subject. Otherwise, it is, within the purpose of these rules, uninsulated.

Irrigation Machine. An electrically driven or controlled machine, with one or more motors, not hand portable, and used primarily to transport and distribute water for agricultural purposes.

Isolated. (as applied to location.) Not readily accessible to persons unless special means for access are used.

Labeled. Equipment or materials to which has been attached a label, symbol, or other identifying mark of an organization that is acceptable to the authority having jurisdiction and concerned with product evaluation, that maintains periodic inspection of production of labeled equipment or materials, and by whose labeling the manufacturer indicates compliance with appropriate standards or performance in a specified manner.

Lighting Outlet. An outlet intended for the direct connection of a lampholder, a luminaire (lighting fixture), or a pendant cord terminating in a lampholder.

Limited Approach Boundary. An approach limit at a distance from an exposed live part within which a shock hazard exists.

Listed. Equipment, materials, or services included in a list published by an organization that is acceptable to the authority having jurisdiction and concerned with evaluation of products or services, that maintains periodic inspection of production of listed equipment or materials or periodic evaluation of services, and whose listing states that the equipment, material, or services either meets appropriate designated standards or has been tested and found suitable for a specified purpose.

Note: The means for identifying listed equipment may vary for each organization concerned with product evaluation, some of which do not recognize equipment as listed unless it is also labeled. Use of the system employed by the listing organization allows the authority having jurisdiction to identify a listed product.

Live Parts. Energized conductive components.

Location, Damp. Locations protected from weather and not subject to saturation with water or other liquids but subject to moderate degrees of moisture. Examples of such locations include partially protected locations under canopies, marquees, roofed open porches, and like locations, and interior locations subject to moderate degrees of moisture, such as some basements, some barns, and some cold-storage warehouses.

Location, Dry.   A location not normally subject to dampness or wetness. A location classified as dry may be temporarily subject to dampness and wetness, as in the case of a building under construction.

Location, Wet.   Installations under ground or in concrete slabs or masonry in direct contact with the earth; in locations subject to saturation with water or other liquids, such as vehicle washing areas; and in unprotected locations exposed to weather.

Medium Voltage Cable.   A single or multiconductor solid dielectric insulated cable rated 2001 volts or higher.

Metal-Clad Cable.   A factory assembly of one or more insulated circuit conductors with or without optical fiber members enclosed in an armor of interlocking metal tape, or a smooth or corrugated metallic sheath.

Metal Wireways.   Sheet metal troughs with hinged or removable covers for housing and protecting electric wires and cable and in which conductors are laid in place after the wireway has been installed as a complete system.

Mineral-Insulated Metal-Sheathed Cable.   A factory assembly of one or more conductors insulated with a highly compressed refractory mineral insulation and enclosed in a liquidtight and gastight continuous copper or alloy steel sheath.

Mobile X-Ray.   X-ray equipment mounted on a permanent base with wheels, casters, or a combination of both to facilitate moving the equipment while completely assembled.

Motor Control Center.   An assembly of one or more enclosed sections having a common power bus and principally containing motor control units.

Nonmetallic-Sheathed Cable.   A factory assembly of two or more insulated conductors having an outer sheath of nonmetallic material.

Nonmetallic Wireways.   Flame-retardant, nonmetallic troughs with removable covers for housing and protecting electric wires and cables in which conductors are laid in place after the wireway has been installed as a complete system.

Open Wiring on Insulators.   An exposed wiring method using cleats, knobs, tubes, and flexible tubing for the protection and support of single insulated conductors run in or on buildings.

Outlet.   A point on the wiring system at which current is taken to supply utilization equipment.

Outline Lighting.   An arrangement of incandescent lamps or electric discharge lighting to outline or call attention to certain features such as the shape of a building or the decoration of a window.

Oven, Wall-Mounted.   An oven for cooking purposes and consisting of one or more heating elements, internal wiring, and built-in or separately mountable controls.

Overcurrent.   Any current in excess of the rated current of equipment or the ampacity of a conductor. It may result from overload, short circuit, or ground fault.

Note: A current in excess of rating may be accommodated by certain equipment and conductors for a given set of conditions. Therefore the rules for overcurrent protection are specific for particular situations.

Overload.   Operation of equipment in excess of normal, full-load rating, or of a conductor in excess of rated ampacity that, when it persists for a sufficient length of time, would cause damage or dangerous overheating. A fault, such as a short circuit or ground fault, is not an overload.

Panelboard.   A single panel or group of panel units designed for assembly in the form of a single panel, including buses and automatic overcurrent devices, and equipped with or without switches for the control of light, heat, or power circuits; designed to be placed in a cabinet or cutout box placed in or against a wall, partition or other support; and accessible only from the front.

Power and Control Tray Cable.   A factory assembly of two or more insulated conductors, with or without associated bare or covered grounding conductors under a nonmetallic jacket, for installation in cable trays, in raceways, or where supported by a messenger wire.

Power-Limited Tray Cable.   Type PLTC nonmetallic-sheathed cable is a factory assembly of two or more insulated conductors under a nonmetallic jacket.

Premises Wiring (System).   That interior and exterior wiring, including power, lighting, control, and signal circuit wiring together with all their associated hardware, fittings, and wiring devices, both permanently and temporarily installed, that extends from the service point or source of power, such as a battery, a solar photovoltaic system, or a generator, transformer, or converter windings, to the outlet(s). Such wiring does not include wiring internal to appliances, luminaries (fixtures), motors, controllers, motor control centers, and similar equipment.

Prohibited Approach Boundary.   An approach limit at a distance from an exposed live part within which work is considered the same as making contact with the live part.

Qualified Person.   One who has the skills and knowledge related to the construction and operation of the equipment and installation and has received safety training on the hazards involved.

Raceway.   An enclosed channel of metal or nonmetallic materials designed expressly for holding wires, cables, or busbars, with additional functions as permitted in this standard. Raceways include, but are not limited to, rigid metal conduit, rigid nonmetallic conduit, intermediate metal conduit, liquidtight flexible conduit, flexible metallic tubing, flexible metal conduit, electrical metallic tubing, electrical nonmetallic tubing, underfloor raceways, cellular concrete floor raceways, cellular metal floor raceways, surface raceways, wireways, and busways.

Receptacle.   A receptacle is a contact device installed at the outlet for the connection of an attachment plug. A single receptacle is a single contact device with no other contact device on the same yoke. A multiple receptacle is two or more contact devices on the same yoke.

Receptacle Outlet.   An outlet where one or more receptacles are installed.

Restricted Approach Boundary.   An approach limit at a distance from an exposed live part within which there is an increased risk of shock, due to electrical arc over combined with inadvertent movement, for personnel working in close proximity to the live part.

Separately Derived System.   A premises wiring system whose power is derived from a battery, from a solar photovoltaic system, or from a generator, transformer, or converter windings, and that has no direct electrical connection, including a solidly connected grounded circuit conductor, to supply conductors originating in another system.

Service.   The conductors and equipment for delivering electric energy from the serving utility to the wiring system of the premises served.

Service Cable.   Service conductors made up in the form of a cable.

Service Conductors.   The conductors from the service point to the service disconnecting means.

Service Drop.   The overhead service conductors from the last pole or other aerial support to and including the splices, if any, connecting to the service-entrance conductors at the building or other structure.

Service-Entrance Cable.   Service-entrance cable is a single conductor or multiconductor assembly provided with or without an overall covering, primarily used for services, and of the following types:

(1) **Type SE.**   Service-entrance cable having a flame-retardant, moisture-resistant covering

(2) **Type USE.**   Service-entrance cable, identified for underground use, having a moisture-resistant covering, but not required to have a flame-retardant covering

Service-Entrance Conductors, Overhead System.   The service conductors between the terminals of the service equipment and a point usually outside the building, clear of building walls, where joined by tap or splice to the service drop.

Service-Entrance Conductors, Underground System.   The service conductors between the terminals of the service equipment and the point of connection to the service lateral.

Note: Where service equipment is located outside the building walls, there may be no service-entrance conductors, or they may be entirely outside the building.

Service Equipment.   The necessary equipment, usually consisting of a circuit breaker(s) or switch(es) and fuse(s), and their accessories, connected to the load end of service conductors to a building or other structure, or an otherwise designated area, and intended to constitute the main control and cutoff of the supply.

Service Point.   The point of connection between the facilities of the serving utility and the premises wiring.

Shock Hazard.   A dangerous condition associated with the possible release of energy caused by contact or approach to live parts.

Show Window.   Any window used or designed to be used for the display of goods or advertising material, whether it is fully or partly enclosed or entirely open at the rear and whether or not it has a platform raised higher than the street floor level.

Signaling Circuit.   Any electric circuit that energizes signaling equipment.

Special Permission.   The written consent of the authority having jurisdiction.

Step Potential.   A ground potential gradient difference that can cause current flow from foot to foot through the body.

Switch, Isolating.   A switch intended for isolating an electric circuit from the source of power. It has no interrupting rating, and it is intended to be operated only after the circuit has been opened by some other means.

Switch, Motor Circuit.   A switch rated in horsepower that is capable of interrupting the maximum operating overload current of a motor of the same horsepower rating as the switch at the rated voltage.

Switchboard.   A large single panel, frame, or assembly of panels on which are mounted on the face, back, or both, switches, overcurrent and other protective devices, buses, and usually instruments. Switchboards are generally accessible from the rear as well as from the front and are not intended to be installed in cabinets.

Touch Potential.   A ground potential gradient difference that can cause current flow from hand to hand or hand to foot through the body.

Unqualified Person.   A person who is not a qualified person.

Utilization Equipment.   Equipment that utilizes electric energy for electronic, electro-mechanical, chemical, heating, lighting, or similar purposes.

Ventilated.   Provided with a means to permit circulation of air sufficient to remove an excess of heat, fumes, or vapors.

Volatile Flammable Liquid.   A flammable liquid having a flash point below 38°C (100°F), or a flammable liquid whose temperature is above its flash point, or a Class II combustible liquid that has a vapor pressure not exceeding 276 kPa (40 psia) at 38°C (100°F) and whose temperature is above its flash point.

Voltage (of a Circuit).   The greatest root-mean-square (rms) (effective) difference of potential between any two conductors of the circuit concerned.

Note: Some systems, such as 3-phase 4-wire, single-phase 3-wire, and 3-wire direct-current, may have various circuits of various voltages.

Voltage, Nominal.   A nominal value assigned to a circuit or system for the purpose of conveniently designating its voltage class (e.g., 120/240 volts, 480Y/277 volts, 600 volts): The actual voltage at which a circuit operates can vary from the nominal within a range that permits satisfactory operation of equipment.

Note: See ANSI C84.1-1995, Voltage Ratings for Electric Power Systems and Equipment (50 Hz).

Voltage to Ground.   For grounded circuits, the voltage between the given conductor and that point or conductor of the circuit that is grounded; for ungrounded circuits, the greatest voltage between the given conductor and any other conductor of the circuit.

Watertight.   Constructed so that moisture will not enter the enclosure under specified test conditions.

Weatherproof.   Constructed or protected so that exposure to the weather will not interfere with successful operation.

Note: Rainproof, raintight, or watertight equipment can fulfill the requirements for weatherproof where varying weather conditions other than wetness, such as snow, ice, dust, or temperature extremes, are not a factor.

Working Near (live parts).   Any activity inside a limited approach boundary.

Working On (live parts).   Coming in contact with live parts with the hands, feet, or other body parts, with tools, probes, or with test equipment, regardless of the personal protective equipment a person is wearing.

II. OVER 600 VOLTS, NOMINAL.

Whereas the preceding definitions are intended to apply wherever the terms are used throughout this standard, the following definitions are applicable only to parts of this standard specifically covering installations and equipment operating at over 600 volts, nominal.

Fuse.   An overcurrent protective device with a circuit-opening fusible part that is heated and severed by the passage of overcurrent through it.

Note: A fuse comprises all the parts that form a unit capable of performing the prescribed functions. It may or may not be the complete device necessary to connect it into an electrical circuit.

Switching Device.   A device designed to close, open, or both, one or more electric circuits.

Circuit Breaker.   A switching device capable of making, carrying, and interrupting currents under normal circuit conditions, and also making, carrying for a specified time, and interrupting currents under specified abnormal circuit conditions, such as those of short circuit.

Cutout.   An assembly of a fuse support with either a fuseholder, fuse carrier, or disconnecting blade. The fuseholder or fuse carrier may include a conducting element (fuse link), or may act as the disconnecting blade by the inclusion of a nonfusible member.

Disconnecting (or Isolating) Switch (Disconnector, Isolator).   A mechanical switching device used for isolating a circuit or equipment from a source of power.

Disconnecting Means.   A device, group of devices, or other means whereby the conductors of a circuit can be disconnected from their source of supply.

Interrupter Switch.   A switch capable of making, carrying, and interrupting specified currents.

Index

About the Authors

■ KENNETH G. MASTRULLO

Kenneth G. Mastrullo, CPE, is a senior electrical specialist in the NFPA Electrical Engineering Department. He serves as a staff liaison to two NFPA committees: NFPA 70E, *Standard for Electrical Safety Requirements for Employee Workplaces,* and NFPA 70B, *Recommended Practice for Electrical Equipment Maintenance.* Ken also is the safety and facilities editor for *necdigest*™. Prior to joining NFPA, he worked more than 28 years in the electrical industry in both the construction and facilities fields as an electrician, project manager, project engineer, and senior facilities engineer. Most recently he was with the Agfa Division of Bayer Corporation (Wilmington, MA), where he was responsible for electrical preventive maintenance, electrical design and implementation, and project management for construction renovation projects at four buildings. Mastrullo is a licensed master electrician in Massachusetts, a certified plant engineer, an OSHA certified outreach trainer, and a member of the AFE, IAEI, IEEE, and IBEW.

■ RAY A. JONES

Ray Jones has spent most of his working life, including 35 years with the DuPont Company, working with industrial electrical systems and installations as a professional electrical engineer. He has served on several national consensus committees and panels, was a charter member of the Petroleum and Chemical Industry Electrical Safety Subcommittee, and is currently chairman of the technical committee of NFPA 70E. The author of many published articles in industry journals, Ray has also presented numerous papers and tutorials on electrical safety issues at technical conferences. Ray coauthored *Electrical Safety in the Workplace*, published by NFPA. He is currently president of Electrical Safety Consulting Services, Inc., a consulting business that helps companies improve their electrical safety programs. Ray is a registered professional engineer.

■ JANE G. JONES

Jane Jones has worked as a book and journal editor, author, newspaper reporter, and technical writer. In addition to serving as a writing consultant and editor for many authors of technical papers, she has authored and edited numerous technical journal articles and has been the technical writer for various codes and standards associated with electrical safety. Along with Ray Jones, she coauthored *Electrical Safety in the Workplace*. Jane is currently vice-president of communications for Electrical Safety Consulting Services, Inc.